SpringerBriefs in Electrical and Computer Engineering

Signal Processing

Series editors

Woon-Seng Gan, Singapore
C.-C. Jay Kuo, Los Angeles, USA
Thomas Fang Zheng, Beijing, China
Mauro Barni, Siena, Italy

For further volumes:
http://www.springer.com/series/11560

Amol D. Rahulkar · Raghunath S. Holambe

Iris Image Recognition

Wavelet Filter-Banks Based Iris Feature Extraction Schemes

Springer

Amol D. Rahulkar
Department of Instrumentation Engineering
AISSMS' Institute of Information
  Technology
Pune, Maharashtra
India

Raghunath S. Holambe
Department of Instrumentation
SGGS Institute of Engineering and
  Technology
Nanded, Maharashtra
India

ISSN 2196-4076   ISSN 2196-4084   (electronic)
ISBN 978-3-319-06766-7   ISBN 978-3-319-06767-4   (eBook)
DOI 10.1007/978-3-319-06767-4
Springer Cham Heidelberg New York Dordrecht London

Library of Congress Control Number: 2014938479

Printed on acid-free paper

Springer is part of Springer Science+Business Media (www.springer.com)

To our families
For their love, encouragement and support

Preface

The increasing demand on enhanced security has led to an unprecedented interest in automated personal recognition based on biometric system. The biometric system makes use of physiological or behavioral characteristics to recognize individuals. Among all the biometric recognition systems, iris recognition system has been deployed in various critical application areas (homeland security, border control, web-based services, national ID cards, etc.) because of its unique, stable, and noninvasive characteristics. It is observed that iris image consists of nonuniform spectral information due to its irregular and random characteristics (tiny crypts, freckles, radial furrows, radial streaks, collarette, pigment spots, etc.).

Multi-resolution analysis (MRA)-based technique can be well suited to represent these iris image structures. It is well known that discrete wavelet transform (DWT) is a powerful tool in MRA. The power of DWT is to offer high temporal localization for high frequencies and good frequency resolution for low frequencies. Most of the iris image representation schemes in the literature used off-the-shelf wavelet basis to extract the features. *Although there is a defined standard for raw iris data, there is none regarding iris feature representation.* Thus, many issues are still open in the field of iris image feature extraction related to the choice of filter bank (FB). The design of FBs and investigations of their properties (near-orthogonality, regularity, time-frequency localization, linear phase, perfect reconstruction, etc.) for image-coding, denoising, compression, etc., have been carried out by many researchers. However, effectiveness of the properties in iris pattern recognition has not been addressed in the literature. Several nonideal factors (eyelids occlusion, multiple/separable eyelashes occlusion, reflection (specular, lighting), poor focus, partially opened eyes, motion blur, noncircular shaped of the iris/pupil, etc.) contained in iris images can increase the false rejection rate (FRR).

This book focused on the design of critically sampled separable and nonseparable wavelet filter banks (FBs) for effective iris image representation. These systems are important for feature extraction algorithms due to their nonredundancy (critical sampling). In addition, *k-out-of-n:A* post-classifier is explored to reduce the FRR. Due to the desired properties of these designed FBs like flexible frequency response, near-orthogonality, and regularity, the filter banks designed in this book can be more effectively used than the existing FBs in many signal processing applications like pattern classification, data-compression, watermarking, denoising, etc. In this book, we have evaluated the performance of the designed FBs for extraction of features of the iris. However, these FBs can be used effectively to extract features from face, fingerprint, palm-print, ear, etc., for automatic person verification (identification).

A brief introduction to biometrics in general and iris in particular is presented in Chap. 1. The motivation along with a brief review of the previously published related work (Iris recognition algorithms, one-dimensional filter-banks, and two-dimensional filter-banks) is also presented in this chapter.

Chapter 2 explains the design of a new class of triplet half-band filter bank (THFB). The properties of the proposed class of THFB are investigated to extract the discriminating iris features. The details of THFB-based feature extraction process including post-classifier are explained in this chapter. We also provide the experimental results to show the effectiveness of the proposed technique.

In Chap. 3, a nonseparable, nonredundant, multiscale combined directional wavelet filter bank (CDWFB) is constructed by the combination of directional wavelet filter bank (DWFB) and rotated directional wavelet filter bank (RDWFB). This chapter also discusses the iris feature extraction algorithm based on a combination of CDWFB and post-classifier. Experimentation is carried out to evaluate the performance of the schemes.

Chapter 4 explains the iris feature extraction scheme based on 2-D nonseparable, nonredundant, multiscale hybrid finer directional wavelet filter bank and classification using fused post-classifier under nonideal environmental conditions. This chapter addresses the issue in the design of DWFB and extends the proposed class of THFB for the 2-D nonseparable filter bank. Simulation results for the proposed algorithm are also presented in this chapter.

Chapter 5 addresses the issue in the design of proposed nonseparable FBs and presents the design of the new class of triplet half-band checkerboard shaped filter bank (THCSFB). This chapter also describes the directional ordinal measures (DOMs) for iris image representation using the designed class of THCSFB. The experimental results are provided to demonstrate the performance of this method.

This book provides the new results in wavelet filter banks-based feature extraction, and the classifier in the field of iris image recognition. It provides the broad treatment on the design of separable, nonseparable wavelet filter banks. It brings together the three strands of research (wavelets, iris image analysis, and classifier). This book contains the compilation of basic material on the design of wavelets that avoids reading many different books. The material on separable and nonseparable wavelet design has been reorganized significantly so to provide an easier path for newcomers and researchers to master the contents.

Acknowledgments

We are grateful to many teachers, colleagues, and researchers, who directly or indirectly helped us in preparing this book. We are thankful to Dr. L. M. Waghmare, Director, Shri Guru Gobind Singhji Institute of Engineering and Technology, Nanded and Dr. P. B. Mane, Principal, Prof. H. P. Chaudhari, Head, Department of Instrumentation Engineering, AISSMS Institute of Information Technology, Pune for their motivation and constant support while preparing the manuscript. We are also thankful to Dr. B. M. Patre, Dr. R. H. Chile, Dr. S. T. Hamde, Dr. V. G. Asutakar, Prof. R. S. Jamkar, Dr. V. S. Thool, Dr. R. V. Sarwadnya, and Mr. R. P. Borse. We would like to acknowledge our colleagues, who have involved indirectly with this work, Dr. D. V. Jadhav, Dr. N. S. Nehe, Dr. M. S. Deshpande, Dr. P. K. Ajmera, Mr. J. P. Gawande, Mrs. S. P. Madhe, and Mr. S. S. Gajbhar. We owe our loving thanks to our staff members Mr. B. D. Upare and Mr. A. J. Shinde for their valuable support. Finally, we wish to acknowledge Dr. Christoph Baumann, Editor, and Sathya Subramaniam for their unusually great help and efforts during the period of preparing the manuscript and producing the book.

Amol D. Rahulkar
Raghunath S. Holambe

Contents

Chapter 1
Introduction

Abstract The brief introduction to biometrics in general and iris in particular is given in Chap. 1. This chapter presents a brief review on iris recognition algorithms, one-dimensional filter-banks, and two-dimensional filter-banks.

Keywords Biometrics · Filter-banks · Iris recognition · Wavelets

A wide variety of systems require reliable person recognition schemes to either confirm or determine the identity of an individual requesting their services. The purpose of such schemes is to ensure that the rendered services are accessed only by legitimate user and no one else. Examples of such applications include secure access to buildings, computer systems, laptops, cellular phones, ATMs etc. In the absence of the robust person authentication schemes, these systems are vulnerable to the wiles of an impostor. Biometric recognition or, simply, biometrics refers to the automatic recognition of individuals based on their physiological and/or behavioral characteristics. By using biometrics, it is possible to confirm or establish an individual's identity based on "who he (she) is?" rather than by "what he (she) possesses?" (e.g., an ID card) or "what he (she) remembers?" (e.g., a password) [1].

1.1 Biometrics

Biometrics described as the science of recognizing an individual based on physiological or behavioral traits has become popular over the traditional token based or knowledge based techniques (i.e. identification cards, passwords, PIN etc.). This is because of the ability of biometrics technology to authenticate an authorized person from an imposter effectively. The problem of resolving the identity of a person can be categorized into two fundamental distinct types of problems known as Verification and Identification.

- Verification (Authentication) system either accepts or rejects the submitted claim of identity. Verification system authenticates a person's identity by comparing the

A. D. Rahulkar and R. S. Holambe, *Iris Image Recognition*,
SpringerBriefs in Signal Processing, DOI: 10.1007/978-3-319-06767-4_1,
© The Author(s) 2014

acquired biometric characteristics with his/her own biometric features stored in the database as reference. In such a system, an individual who desires to be recognized claims an identity, usually via a person identification number (PIN), a user name and biometric. System conducts one-to-one comparison to determine whether the identity claimed by an individual is true.

- Identification system recognizes an individual by searching the entire database (of characteristics of authenticated users) for the closest match. It conducts one-to-many comparisons to establish the identity of an individual. In an identification system, the system establishes a subject's identity without claim of its identity (Who am I?).

1.2 Requirement of Biometrics Systems

Any human physiological or behavioral characteristics can be used to make person authentication as long as it satisfies the following requirements: [2, 3].

- Universality: All human beings are endowed with the same physical characteristics—such as fingers, iris, face, etc. which can be used for identification.
- Uniqueness: The physiological or behavioral characteristics are unique for each person, and thus constitute a distinguishing features.
- Permanence: These characteristics remain largely unchanged throughout a person's life.
- Collectability: A person's unique physical characteristics need to be acquired in a reasonably easy way for quick identification.
- Performance: The degree of accuracy of identification must be quite high before the system can be operational.
- Acceptability: Applications will not be successful if the public offers strong and continuous resistance to biometrics.
- Circumvention: In order to provide added security, a system needs to be harder to circumvent than existing identity management systems.

It is necessary that a practical biometric system should have high recognition accuracy and computational speed. It should have wide acceptability by the people and should be robust to various fraudulent methods [4]. A number of biometric authentication systems have been in use in various applications. Each one has its strengths and weaknesses and its choice is typically depends on the application.

1.3 Iris as a Biometric

The main purpose of biometrics is to identify an individual based on physiological and/or behavioral attributes such as fingerprint, face, hand geometry, speaker, retina, ear, iris, palmprint, voice, signature, gait etc. as shown in Fig. 1.1. Biometric recognition system first captures the data, removes the redundant information from data (extracts relevant features) and constructs a template from it. Then, compares this

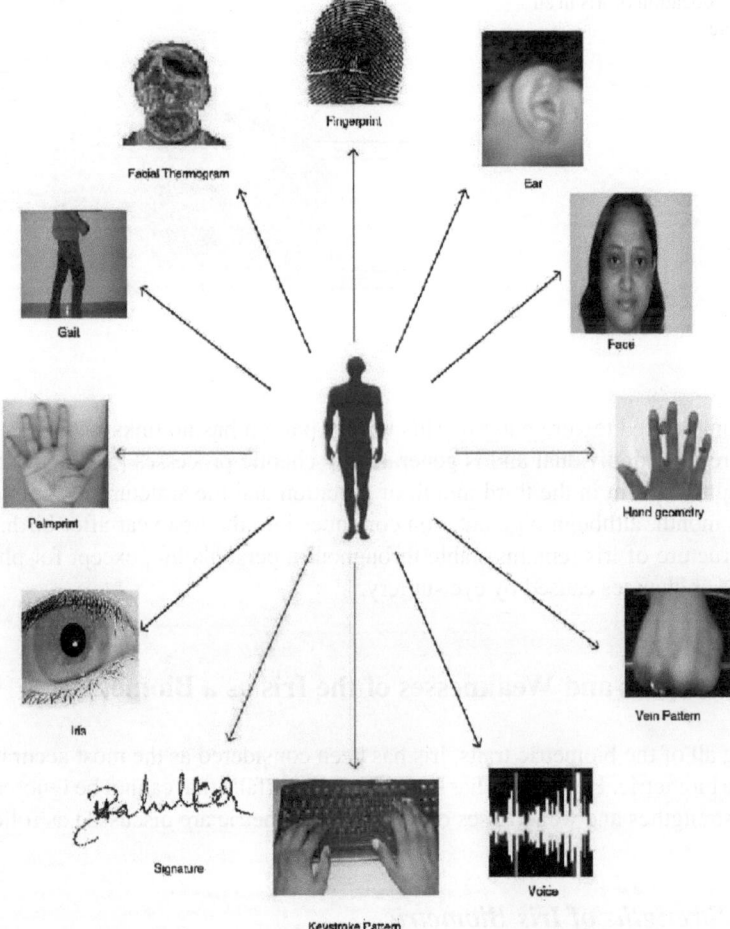

Fig. 1.1 Examples of biometric traits that can be used for authenticating an individual

template from a database of such templates in order to verify or identify the person. Among all the biometric recognition systems, iris recognition system has been deployed in various critical application areas (homeland security, border control, web-based services, national ID cards, rapid processing of passengers, missing child identification, welfare distribution etc.) because of its unique, stable and non-invasive characteristics [5–7].

Iris is an annular part between the pupil (black portion) and sclera (white portion) of an eye image as shown in Fig. 1.2. The iris is the only internal organ of the body that is readily visible from outside. Its purpose is to control the amount of light that enters into the eye through the pupil, by using dilator and sphincter muscles to control the pupil size. It is made up of an elastic fibrous tissue that gives it a

Fig. 1.2 Location of iris in an
eye image

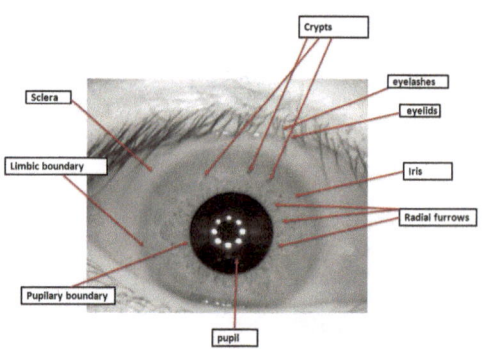

very complicated texture pattern. This texture pattern has no links with the genetic
structure of an individual and is generated by chaotic processes [5–7]. The human
iris begins to form in the third month of gestation and the structure is complete by
the 8th month, although pigmentation continues into the first year after birth. After
that, structure of iris remains stable throughout a person's life, except for physical
damage or changes caused by eye surgery.

1.4 Strengths and Weaknesses of the Iris as a Biometric

Among all of the biometric traits, iris has been considered as the most accurate and
reliable biometric. However, it has some intrinsic pitfalls that cannot be ignored. The
major strengthes and weaknesses of the iris as biometric are discussed as follows:

1.4.1 Strengths of Iris Biometric

- Iris pattern has small intra-class variability.
- The iris is well protected and is an internal organ of the eye. It contains high degree
  of randomness [5, 6, 8].
- Iris is externally visible and the iris image acquisition is possible from a distance.
- The iris pattern remains stable throughout the lifetime of a person and it is assumed
  that each individual has a unique iris pattern [5–7].
- No evidence of genetic influence has been found in the structure of the iris. Hence,
  the iris structures in both eyes of the same person are different and those of identical
  twins are also different [5–7].

1.4.2 Weaknesses of Iris Biometric

- It is difficult to capture the iris image since the size of the iris is very small (its approximate diameter is 1 cm). A specialized camera with an extensive apparatus setup is needed to acquire the iris images.
- The iris could be partially occluded by lower and upper eyelids, and obscured by eyelashes, reflections, lenses etc.
- There is non-elastic deformation due to the dilation and contraction of the pupil.

1.5 Performance Measures of Iris Recognition System

The degree of similarity between two biometric feature sets is indicated by a similarity score. A similarity match score is known as a genuine or authentic score if it is a result of matching two samples of the same biometric trait of a user. It is known as an impostor score if it involves the comparison of two biometric samples originating from different users. An impostor score that falls below the threshold η results in a false accept or false match, while a genuine score that exceeds the threshold η results in a false reject or false non-match. The False Acceptance Rate (FAR) or False Match Rate (FMR) of a biometric system can therefore be defined as the fraction of impostor scores falling below the threshold η. Similarly, the False Rejection Rate (FRR) or the False Nonmatch Rate (FNMR) of a system defined as the fraction of genuine scores exceeding the threshold η. When a linear, logarithmic or semi-logarithmic scale is used to plot these error rates i.e. FAR versus FRR, the resulting graph is known as a Receiver Operating Characteristic (ROC) curve. The performance of a biometric system can also be summarized using other single-valued measure such as the equal error rate (EER). The EER refers to that point in a ROC curve where FAR equals the FRR.

1.6 Nonideal Iris Recognition: A New Challenge

Most of the iris recognition algorithms are effective in a cooperative environment, with an inflexible image capturing setup and exhibit a very high recognition accuracy. However, their performance affects when the iris images captured in an unconstrained situation and contain occlusion of eyelids, eyelashes, shadows, specular reflection etc. Moreover, the inner and outer boundaries of the iris may not maintain any particular shapes and therefore, the segmentation performance will be decreased if those boundaries are not accurately detected and modeled using a more flexible and generalized segmentation method. Other challenges include defocusing, motion blur, poor contrast, lighting reflections, camera diffusion, head rotation, gaze direction, camera angle, pupil dilation etc. Thus, the nonideal conditions contained in iris images affect the overall recognition accuracy. Hence, it is desirable to design the robust iris recognition method that can cope with such artifacts and thereby increase the iris recognition accuracy. Some of the common artifacts present in iris images are discussed as follows:

1. **Eyelid Occlusions**: The upper and lower eyelids may occlude the significant portion of the iris, especially in its extreme vertical position. The eyelid occlusion can also occur in the lower portion of the normalized iris images.
2. **Eyelash Occlusions**: Eyelashes are divided into two types, namely: The separable eyelashes and multiple eyelashes [8]. Separable eyelashes are isolated in the eye image, and multiple eyelashes are bunched together and overlapped in the eye image.
3. **Specular Reflections**: This type of reflection may occur within the iris region and may corrupt significant portion of the iris pattern. The specular reflections are usually caused by imaging under natural light.
4. **Lighting Reflections**: This type of reflections may occur due to artificial light sources that are close to the subject, and also under natural lighting conditions. These reflections are highly heterogeneous in nature as they can appear with a broad range of dimensions and can be located in different regions of the iris [9]. The occluded areas of the iris can typically have relatively higher intensity values than the areas affected by specular reflections.
5. **Poor Focus**: The poor focus has been a critical issue for accurate iris segmentation. This may occur due to moving objects that may interact with the camera in the less constrained capturing setup and to the limited depth-of-field of any imaging system [9]. A small deviation in the distance between the eye and the camera position may create severe focus problems and this subsequently increases the FRR.
6. **Completely Closed and Partially Opened Eyes**: The completely closed eye is a severe noise factor that prevents any kind of recognition. If the eyes are partially opened, signification portions of the iris cannot be extracted due to occlusion, which further affects the segmentation performance.
7. **Motion Blur**: The iris images captured in a noncooperative imagery setup, where the subject or camera is moving, may be affected by motion blur.
8. **Nonelliptical/Noncircular Shapes of the Iris/Pupil**: The pupillary and limbic boundaries often have arbitrary shapes. Therefore, if these boundaries are fitted with some simple shape assumptions, then this can lead to inaccurate segmentation results. Figure 1.3 shows some of these artifacts.

1.7　A Brief Review On

The literature has been reviewed and classified in the following categories:

- Iris Recognition Algorithms
- Two-channel One-dimensional Filter Banks and
- Two-dimensional Filter Banks

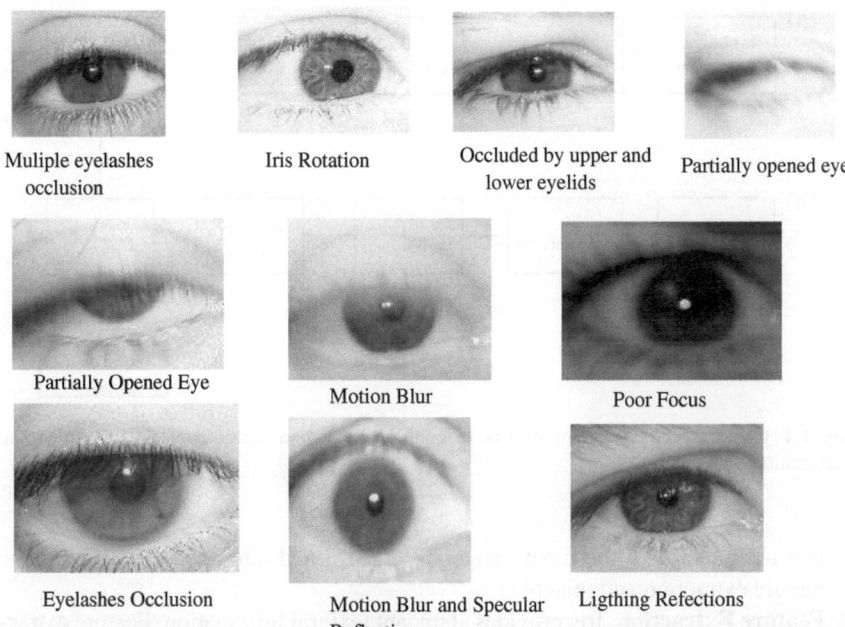

Muliple eyelashes occlusion   Iris Rotation   Occluded by upper and lower eyelids   Partially opened eye

Partially Opened Eye   Motion Blur   Poor Focus

Eyelashes Occlusion   Motion Blur and Specular Reflection   Ligthing Refections

Fig. 1.3 Some of the artifacts present on iris images

1.7.1 Iris Recognition Algorithms

Since 1985, many researchers have worked on the problem of identifying an individual from iris pattern. It has been discovered that every iris is unique to each subject, particularly in the detailed structure of front or anterior layer. Iris of identical twins are not only different, but the iris of two eyes of the same person are also different. The generalized block diagram of iris recognition system in shown in Fig. 1.4. The system is divided into enrollment and authentication modules. The enrollment process consists of iris acquisition, segmentation, normalization, and extraction of features from the iris images. These features are stored in the database as reference. During the recognition (decision) process, the test iris feature is compared with stored features resulting in an acceptance or rejection decision of the input iris pattern. Thus, iris recognition process mainly divides into four important steps:

1. **Iris Segmentation**: The first step of any iris recognition system is to locate the iris rim defined by the inner (pupillary boundary) and outer boundaries (limbic boundary) of iris.
2. **Normalization**: Iris of different people may vary in size. Even for the same person, size of the iris may vary due to contraction and dilation of pupil caused by the variation in illumination and other factors. Normalization process ensures

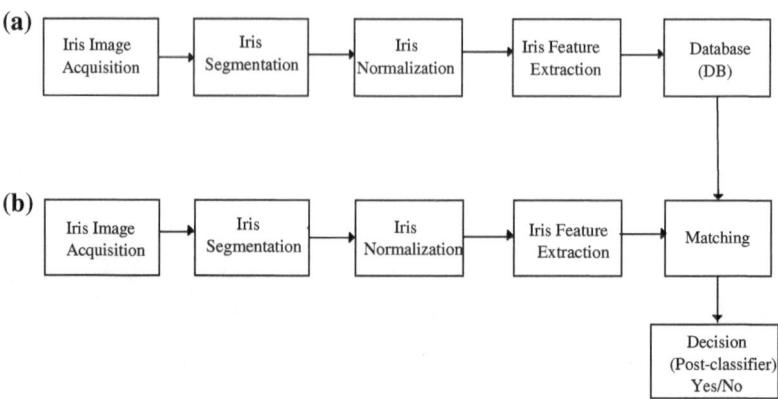

Fig. 1.4 General block diagram of iris recognition system **a** Enrollment **b** Authentication/ Recognition

that iris patterns of the input image are transformed into fixed size to facilitate feature extraction and matching.

3. **Feature Extraction**: Iris provides abundant textural information. Feature extraction algorithm encode this textural pattern to form a feature vector for the comparison.

4. **Matching**: Feature vectors are classified through different matching algorithms such as Euclidean distance, Hamming distance, Canbera distance etc.

Daugman [5–7] proposed the first successful iris recognition algorithm that has been widely used in commercial iris recognition products. A digital video camera is employed to capture several sequential images of an iris in the near infrared (NIR) frequency range, usually with the subject's co-operation to center the eye in the view of the camera via a feedback video loop. NIR is used because it provides abundant textural information of iris, so more useful features can be extracted. While the images are being acquired, the image quality is assessed by measuring the spectral power in the mid to high frequency portion of the two dimensional (2-D) Fourier transform (FT). This quantity is then maximized by either adjusting the camera or indicating to the subject a need for adjustment. During the preprocessing stage, integro-differential operator (IDO) is developed to detect the inner and outer boundaries of iris. A similar operation is used to locate the upper and lower eyelid boundaries. This segmented circular iris ring is converted into rectangular block of fixed size using polar transformation commonly known as a Daugman's homogeneous rubber sheet model (DRSM). A series of modified complex valued 2-D Gabor wavelets are applied on the local regions of the normalized iris image in order to extract the phase information of the resultant iris image. This phase information of local iris texture is coarsely quantized into 2-bits binary codes so as to generate 256-byte IrisCode. The Hamming distance (HD) is used to find the probability of disagreement between the two IrisCodes.

Wildes [10] used gradient-based Hough transform (HT) to decide the two circular boundaries of an iris. This system also models the eyelids as parabolic arcs and detected the upper and lower eyelids using HT. It is observed that the distinctive spatial characteristics of the human iris provided better visibility at different scales i.e. distinguishing structure range from the overall shape of iris to the distribution of tiny crypts and detailed texture. In order to capture such range of spatial details, isotropic bandpass decomposition is derived from the application of Laplacian of Gaussian (LoG) and is used to iris images. The matching process is based on the normalized correlation between the acquired and database representation. In addition, Fisher linear discriminant (FLD) is used to either accept or reject the identity of a person. Boles and Boashash [11] proposed iris recognition algorithm based on the zero-crossing of the wavelet transform (WT) [12]. They have localized and normalized the iris using edge detection and other well known computer vision algorithms. Zero-crossings of the WT (using first derivative of cubic spline function) are computed at various resolutions over concentric circles on the iris image. The resultant one-dimensional (1-D) signals are compared with model features using four different dissimilarity functions. Sanchez-Avila et al. [13] presented the same type of algorithm based on the dyadic discrete wavelet transform (DWT). Masek [14] used circular HT to detect the iris/pupil boundaries. The 1-D log-Gabor filters were employed to extract the iris features and HD was used for template matching.

Ma et al. [15] used the two features (first is total spectrum power of an iris region and second is the ratio of middle frequency power to other frequency power of iris image) in order to assess the quality of iris images. These two features are used to train support vector machine (SVM) so as to predict the quality of the input iris images. The good quality iris images are localized by the combination of Canny edge operator and HT. The localized iris ring is unwrapped to a rectangular block with fixed size. This normalized iris image has low contrast and may have non-uniform illumination caused by the position of light sources. These may affect the subsequent processing in feature extraction and matching algorithms. In order to obtain a more well-distributed texture image, iris image has been enhanced by background-subtraction method. A bank of spatial filters is constructed to capture the local details of iris. The feature vector (FV) is derived by computing the mean and average absolute deviation of each 8×8 small block of the filtered iris images. To improve the computational efficiency and classification accuracy, FLD is used to reduce the dimensionality of the feature vector and then the nearest center classifier is adopted for classification. The same authors presented an efficient algorithm for iris recognition by characterizing key local variations [16]. The local sharp variation points of iris (denoting the appearing or vanishing of an important iris image structure) are utilized to represent iris characteristics. Firstly, a set of 1-D intensity signals is constructed to characterize the most important information from the original 2-D iris image effectively. Then a special class of 1-D wavelet (quadratic spline of a finite support) is used to represent the resulting 1-D intensity signals. The position of local sharp variation points is recorded as binary features. The similarity between a pair of feature vectors is calculated using Hamming distance. Ma et al. [17] constructed 1-D signal that contains majority of local variations of the iris. Gaussian-Hermite moments of such intensity signals are used as distinguishing features. The same process (mentioned in [15]) is carried out for dimensionality reduction and classification of iris. In their next work [18], multi-channel Gabor filter-based iris recognition scheme has been proposed. The circular symmetry filters were used by Ma et al. [19] for iris recognition on the enhanced normalized iris images. Sun et al. [20] presented an elastic blob matching algorithm to overcome the limitations of local feature based classifiers (LFC). In addition, cascade classifier scheme is proposed to combine LFC and an iris blob matcher. Sun and Tan [21] proposed ordinal measures (OMs) for iris feature representation in order to characterize qualitative relationships between the iris regions rather than precise measurements of iris image structures. They have preprocessed the original iris image as given in [15] and [17]. In their work, multilobe differential filters (MLDFs) based on 2-D Gaussian filter have been presented for ordinal iris feature extraction. These ordinal filters are used on 1024 densely sampled image regions to obtain 128 bytes ordinal code for every iris image with flexible interlobe distance d. The error rate has been estimated using bootstrap method on the measured Hamming distances between two ordinal templates of the same class. Dong et al. [22] presented a personalized iris matching strategy using a class-specific weight map learned from the training images of the same iris class. Iris image preprocessing includes iris localization and normalization. Adaboost-cascade iris detector is used to determine the rough position of iris center, and then apply an elastic model named "pulling and pushing" to refine edge of iris and pupil, and finally remove the eyelash and shadows via prediction model [23]. The authors have used same iris feature representation scheme as given in [21] and the weight map of each iris class is learned based on intra-class Hamming distances among many registered templates. This weight map updated and stabilized by increasing the number of training images.

Vatsa et al. [24] proposed algorithms for iris segmentation, quality enhancement, match score fusion, and indexing to improve the accuracy and speed of iris recognition. In their work, a two-level hierarchical iris segmentation algorithm is proposed to detect the iris boundaries from nonideal iris images. In the first level, intensity thresholding is used to detect an approximate elliptical boundary and Mumford-Shah functional is used at the second level to obtain the accurate iris boundary. A support vector machine (SVM) based iris quality enhancement algorithm is described to obtain a high-quality feature-rich iris image. The textural features are extracted by 1-D log-Gabor transform that is invariant to translation and rotation, and the topological features are extracted using Euler number scheme which is invariant under translation, rotation, scaling, and polar transformation. Then, 2ν-SVM is proposed to develop a fusion algorithm that combines the match scores obtained by matching textural and topological features. Schuckers et al. [25] presented a new method for the compensation of off-angle occluded iris images in order to improve iris recognition performance. Two approaches are proposed to account for angular variations in the iris images. In the first approach, Daugman's IDO is used to estimate the gaze direction. After the angle is estimated, the geometric transformation is incorporated on off-angle iris image to obtain frontal view iris image. This frontal view iris image normalized using DRSM and enhanced by standard histogram equalization method to boost the intensity of poorly visible regions. The global independent component analysis (ICA) is applied on this enhanced normalized iris image to encode the iris. Euclidean distance measure is used to perform matching between two projected iris images. The second approach is based on angular deformation calibration model to compensate off-angle to create frontal view iris image. The normalized iris information is transformed into the wavelet domain using the biorthogonal 5/3 wavelet basis functions that are designed using the lifting steps. Hamming distance has been used for the matching. Abhyankar and Schuckers [26] introduced biorthogonal wavelet neural network (BWN) for off-angle iris recognition by adjusting non-ideal factors through repositioning the BWN.

Velisavljevic [27] presented iris coding and recognition based on directionlets where iris is localized using IDO with few modifications. The oriented separable wavelet filters (*directionlets*) [28] are used to provide directional iris features along both radial and angular directions. The resulting transform coefficients are sampled to obtain the binary code and these codes are compared using weighted HD. Monro et al. [29] presented an iris coding scheme based on the difference of discrete cosine transform (DCT) coefficients of overlapped angular patches from normalized iris images. In their work, good quality iris images have been selected based on their kurtosis values and localized the iris by Hough transform. This localized iris image is normalized by DRSM and enhanced using background-subtraction method [15]. The DCT of a series of averaged overlapping angular patches are computed from enhanced normalized iris image and a small subset of coefficients is used to form subfeature vectors. Iris code is generated as a sequence of many such subfeatures, and the classification is carried out with the help of weighted Hamming distance. Proenca and Alexandre [30] partitioned a normalized iris image into six regions and obtained six IrisCodes using Gabor filters. The matching scores of these six regions are fused together to generate an overall matching score for the iris recognition. Huang et al. [31] proposed a rotation invariant iris recognition scheme based on non-separable wavelets. Firstly, a bank of non-separable orthogonal wavelet filters is used to capture the iris characteristics. Secondly, a method of Markov random field is used to capture rotation invariant iris features. Finally, two-class kernel Fisher classifiers are adopted for classification.

Park et al. [32] presented filterbank-based iris recognition scheme that extract the directional features of iris on multiple scales. This scheme first localized the iris using IDO, normalized by DRSM, and established a region of interest (ROI) for feature extraction. Secondly, iris features are extracted on multiple scales from the ROI and constructed the feature vector using the directional filter bank (DFB) designed in [33]. Iris image is decomposed into various directional subbands and Hamming distance is used to find the dissimilarity between two iris images. Nabti et al. [34] introduced a multiscale approach for edge detection using wavelet maxima modulus. Eyelids and eyelashes are isolated from the detected iris image using linear HT. This segmented circular iris ring is unwrapped into rectangular region using DRSM. The combined multiresolution iris feature extraction scheme is presented by analyzing the iris using wavelet maxima components (given in [12]) and then applying a special class of Gabor filter bank on the normalized image. Two types of binary feature vectors have been computed using statistical measures (mean and variance, and moment invariance) and Hamming distance is used to verify the two iris samples.

Thornton et al. [35] carried out the same type of experimentation as used in [5] for segmentation, normalization, and feature extraction. Then, they have presented a general probabilistic framework (Bayesian approach) to deformed pattern matching of iris images. A maximum posteriori probability (MAP) estimate of the relative distance between a pair of iris images has been derived. Tisse et al. [36] introduced a combination of IDO with HT to localize the iris and DRSM is used to normalize the iris. The iris characteristics have been analyzed using analytic image that is constructed from the original image and its Hilbert transform. The emergent frequency functions are sampled to form a binary feature vector and used Hamming distance for matching. Lim et al. [37] used the same technique given in [15] to localize and normalize the iris. In their work, iris image decomposed into four levels using 2-D Haar WT and quantized the fourth-level high-frequency information to form an 87-bit iris code. A modified competitive learning neural network (LVQ) was adopted for classification. Helen and Selvan [38] used contourlet transform to extract the iris texture. Altan [39] used first generation curvelet transform to enhance (denoise) iris images (also fingerprint images) and 2-D Gabor wavelet for feature extraction followed by genetic algorithm (GA) for dimensionality reduction. Chen et al. [40] presented wavelet-based quality measures for iris images and suggested that different regions of the iris have different qualities and local iris image regions with better quality have better classification capability and vice-versa. Du et al. [41] segmented the iris by the combination of Canny edge operator and Sobel operator and normalized the iris using resolution-invariant polar coordinates (which is different from DRSM). The grayscale-invariant local texture patterns (LTPs) are developed to extract the local iris features. In addition, distance measure is developed named as *Du measure* to measure the similarity between a test iris signature and the signatures in the database. In [42], multialgorithmic fusion has been presented for iris recognition. This method combined two types of iris recognition algorithms: one is based on phase information [5] and the other is based on zero-crossing representation [11]. These two algorithms fused at the matching score and a fusion scheme is developed to generate a fused score that is used to take the final decision. Ariyapreechakul and Covavisaruch [43] presented partial iris recognition system based on Radon transform. Ali and Hassanien used Haar WT for iris feature extraction. Sung et al. [44] proposed the scheme to localize the iris area between the inner boundary and collarette boundary in order to remove unnecessary area and to increase the recognition rate. Discrete wavelet transform (db4) is used to extract these iris features and recognize the person using SVM. Ajay and Passi [45] presented a comparative study of iris recognition performance using log-Gabor, Haar wavelet, DCT, and FFT based features. Pourbseri and Araabi [46] has been utilized the lower part of the enhanced iris area to extract the features using DWT (db2). The mixed classifier is exploited using Hamming distance and harmonic mean distance for the recognition purpose. Dey and Samanta [47] presented an approach for feature extraction and matching process. DWT (db4) with four levels is used to extract the iris features and quantized these features into 4 quantization levels. In addition, different weights have

Fig. 1.5 Block diagram of
two-channel PRFB

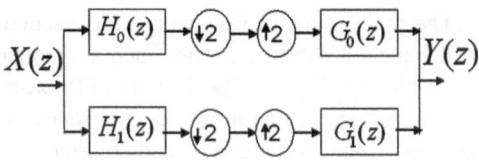

been assigned to different iris regions to compare the two iris templates. Bodade and Talbar [48] suggested the use of complex wavelet filters and rotated complex wavelet filters (RCWFs) for iris feature extraction.

Noh et al. [49] presented an iris recognition system based on independent component analysis (ICA) to generate optimal basis vectors. Dorairaj et al. [50] presented a global iris encoding scheme using ICA to improve the performance. Krichen et al. [51] suggested phase-correlation-based iris matching approach to deal with degradations in iris images that occur due to unconstrained acquisition procedures. This matching scheme is a fusion of global and local Gabor based phase-correlation schemes. Chou et al. [52] suggested non-orthogonal view iris recognition system that consists of iris imaging, iris segmentation, iris feature extraction, and a classification modules. In their work, an intelligent and robust random sample consensus iris segmentation method is proposed to detect iris boundaries in four-spectral iris image. Next, an edge-type iris feature descriptor has been proposed in the same paper. The descriptor characterized an iris pattern with multiscale edge-type maps detected using derivative of Gaussian (DoG) and Laplacian of Gaussian (LoG) filters. The matching of two iris patterns has been formulated with the help of classifier ensembles. A detailed literature survey of iris recognition algorithms up to the year 2007 can be found in [53].

1.7.2 Two-Channel One-Dimensional Filter Banks

Wavelets and filter banks are widely used in digital signal/image processing applications. In a conventional filter bank, a signal is first decomposed into various bands of frequencies for various applications. These might include subband coding wherein each subband is further decomposed and then subsequently coded. In image processing, wavelets are used for feature extraction which forms an important intermediate step in computer vision. Each application requires a specific kind of filter bank. Filter banks differ in the design of the constituent filters. A perfect reconstruction filter bank (PRFB) is one where a signal is first decomposed into various subbands for the purposes of an application and then combined together to reconstruct the original signal perfectly. Figure 1.5 shows a block diagram of a PRFB. This FB is characterized by its analysis (H_0 and H_1) and synthesis (G_0 and G_1) filters. H_0 and G_0 aim to be ideal lowpass filters with cutoff frequency of $\frac{\pi}{2}$ and H_1, G_1 aim to be ideal highpass filters with cutoff frequency of $\frac{\pi}{2}$.

The main focus in FB theory and practice is to design critically sampled FBs with some desirable properties such as frequency selectivity, regularity, symmetry, near-orthogonality etc. The design of FBs with these properties is a challenging task and hence received an attention over many years. Conventional filter bank design revolves around "orthogonal" and "biorthogonal" FBs. There exists different design techniques for these two types of filter banks. Although, both these types have their respective advantages, the biorthogonal type is preferred in image processing applications because the associated filters provide linear phase that is desirable in these applications. The idea of perfect reconstruction (PR) is first given by Brune [54]. This has been applied to the design of FBs by Swamy and Rajan [55].

The initial concern in the FB design was focused on alias cancellation. Croisier et al. [56] addressed this aliasing problem by quadrature mirror filter (QMF) solution in the two-channel case and became popular both in speech [57] and image [58] coding applications. The aliasing cancellation problem was then solved in the general case for an arbitrary number of channels [59–62]. Following this, efforts were directed towards addressing the problem of perfect reconstruction. The initial solution for the two channel case was found by Smith and Barnwell [60] and was followed by solutions for an arbitrary number of channels [61, 62]. A time domain view of some of these results was given by Nayebi et al. in [63]. An excellent overview of the subject is given by Vaidyanathan [64].

Various methods for the design of 1-D linear phase two channel PRFBs have been proposed. One of the most common design methods for designing wavelet filters is the factorization of a halfband polynomial [65–67]. In these approaches, a halfband product filter $P(z)$ is first designed, and is then factored into two filters, yielding the analysis and synthesis lowpass filters in the FB. Most of the biorthogonal FBs are designed by the factorization of halfband polynomial (HBP) which has maximum number of zeros at $z = -1$. It is well known that most of the popular biorthogonal FBs e.g. Cohen-Daubechies-Feauveau (CDF-9/7), spline family of wavelet FBs [68–70] are designed by the factorization of a Lagrange half-band polynomial (LHBP). The LHBP has maximum number of zeros at $z = -1$ to achieve the maximum regularity. However, LHBP filters do not have any degree of freedom and thus there is no direct control over frequency response of the filters. In order to have some independent parameters (which can be optimized to obtain some control over frequency response of the filter), Patil et al. [67] used general half-band filter factorization (not LHBP) to design two-channel bi-orthogonal wavelet FIR FBs (BWFB).

A new construction of wavelets, called the lifting scheme, has been developed by Daubechies and Sweldens [71, 72]. Lifting structure is one of the alternative schemes to design FBs which provide structurally imposed PR property. The advantage of this method is that it is a more efficient implementation than the classical WT. Lifting makes optimal use of the similarities between the lowpass and the highpass filters to speed up the calculation. The inverse WT can be processed immediately by reversing the operations of the forward transform.

Based on the lifting scheme, Phoong et al. [73] introduced a class of halfband pair filterbank (HPFB) structure based on two kernels. In HPFB structure, magnitudes of the frequency responses of the analysis and synthesis low pass filters at

$\omega = \pi/2$ are restricted to 0.5 and 1.0 respectively or vice-versa. This encounters certain restrictions for the control of its frequency response. In order to overcome this restriction [73], Ansari et al. [74] proposed the FB structure based on a class of triplet halfband filterbank (THFB) using three kernels that has structural PR, feature-rich structure and simple design. In their work, two methods are proposed to design two-channel 1-D biorthogonal FBs based on the triplet of half-band filters. The first method is based on Lagrange filter coefficients to achieve maximal flatness in the filters (due to maximum regularity) with slow frequency roll-off. The second method is employed Remez algorithm which results in equiripple filters with user-defined cut-off frequencies. It has sharp frequency roll-off but does not have the regularity condition. In their work, the shape parameter p is used to achieve a greater flexibility in frequency response of the filters. However, regularity order related to the number of zeros at $z = -1$ has not been specified. In order to bridge the gap between these two extremes, Tay and Palaniswami [75] introduced a novel approach to design a class of THFB. Parametric Bernstein polynomial is used to impose the required vanishing moments (VMs). This technique is based on iterative least square approach to determine the coefficients of Bernstein polynomial. However, Bernstein polynomial is suitable for nearly maximally flat frequency response rather than for ripple responses with sharp roll-off. In [76], Tay modified the work of [75] and suggested a new class of even-length biorthogonal wavelet filters for Hilbert pair design (ETHFB). Chan and Yeung [77] presented a design of THFB with regularity using semi-definite programming (SDP). However, the semi-infinite constraints are approximated by a large finite number of constraints that give an inefficient design for higher order filters. Kha et al. [78] proposed an efficient SDP method in order to design a class of THFB with optimal frequency selectivity for a given regularity condition. Eslami and Radha [79] generalized Ansari's method to a multidimensional FB design with any number of analysis and synthesis VMs using a structural approach based on Kovacevic method [80]. The filters designed in [79] achieved better regularity, lower frame bounds and better frequency selectivity than the filters designed in [80].

Various time domain optimization approaches have been presented in [63, 81]. In these approaches the filters in the FB are designed by the 'time domain' optimization of the filter coefficients (i.e. a direct optimization of the filter coefficients, without any parametrization), after imposing the PR constraints on the coefficients.

1.7.3 Two-Dimensional Filter Banks

The application of wavelets to image processing requires the design of 2-D wavelet bases. The traditional way to obtain 2-D wavelets is to apply 1-D filters separately in horizontal and vertical directions. This separable FBs provide only three directional information (horizontal, vertical, and mixed diagonal). It also suffers from poor diagonal orientation selectivity as 45° and 135° directions are combined into one subband in each resolution. This is because such separable 2-D filters have the limitations that the frequency support of the filters in separable 2-D FB is always in rectangle shape. However, since the human visual system (HVS) is less sensitive to high frequencies in both directions (horizontal and vertical) simultaneously, a non-rectangular frequency region is more desirable. Such a frequency region (with non-rectangular frequency region) can only be obtained with non-separable FBs. These non-separable FBs are used to provide more directional information.

A common approach used in the design of 2-D FBs is the McClellan transformation. This consists of "transformation of variable" approach to design 2-D filters. Nonseparable filters are better suited for the representation of multidimensional (MD) signals. Viscito and Allebach [82] introduced the key notations and concepts for the development of the theory of MD filter banks using the concept of lattices. They discussed constraints on the frequency supports of the filters in a nonseparable critically sampled FB, and also demonstrated that different frequency supports with various shapes are possible when designing a critically sampled nonseparable FB using a given sampling matrix.

The design of general multidimensional FBs is a challenging task. The design of critically sampled FBs leads to a factorization problem (to factor a given polynomial into non-trivial polynomial factors). This problem is addressed in the 1-D case by the fundamental theorem of algebra which states that any 1-D polynomial can be factored into products of degree one polynomials. However, due to lack of fundamental theorem of algebra for two-dimensional polynomial factorization, design of 2-D FBs becomes more challenging. Along similar lines as the cascade structures proposed by Vaidyanathan for 1-D FBs [83], and then further researched by a number of authors for the 1-D case [65, 83, 84], cascade structures for the design of PR 2-D nonseparable FBs were proposed in [80, 85], and [86]. These structures are based on the 1-D structures, so that multidimensional FBs can be designed in a similar way as the 1-D case. However, there are no completeness in the design of non-separable multidimensional filter banks [83]. Nonseparable FBs as tools for the directional decomposition of images were first described by Bamberger and Smith [33], and then further investigated by Hong and Smith [87], and Nguyen and Oraintara [88].

The directional filter bank (DFB) [33] has been widely used as a basic building block in the development of nonseparable expansions for the directional representation of signals. A group theoretic approach to MD FBs has been discussed by Kalker and Shah [89]. Use of the Cayley Transform [90] in the characterization and design of paraunitary multidimensional FBs has been discussed by Zhou et. al [91, 92]. Design of modulated 2-D FBs has been presented by Lin and Vaidyanathan [93]. An important advantage of the modulated FBs is that only a single prototype filter needs to be designed. Various shapes of the prototype filter have been considered. Ikehara [94] considers prototype filters with triangle shaped passband support, while Lin and Vaidyanathan [93] consider the design of cosine-modulated 2-D FBs with a parallelogram shaped prototype filter. The FBs in [93] are referred to as two-parallelogram filter banks. The passband support of each filter in the two-parallelogram FBs is a union of two-parallelograms and hence called as "two-parallelogram FBs".

The two-channel Quincunx FBs have received particular attention. In 2-D, Quincunx lattice is the only nonseparable lattice with a subsampling factor of two.

The most commonly used method for the design of the Quincunx FBs is the method of variable transformations as described by Shah and Kalker in [95], and Tay and Kingsbury in [96]. In this method, a 1D prototype two-channel FB is designed first. Then it is mapped to a multidimensional FB by a change of variables using the McClellan transformation [97]. A design method where a transformation is applied to the polyphase components of the filter instead of the original filter transfer function has been described by Phoong et. al [73]. The transformation based designs have the restriction that one cannot explicitly control the shape of the filter-responses of the filters in the FB. Also, the method of transformation cannot be generalized to the design of FBs with general nonseparable sampling matrices other than the Quincunx matrix.

There are a number of approaches available in the literature to provide more directional information using non-separable FBs that include directionlets [27], hybrid wavelets and DFBs [98], multidimensional filter bank using triplet of Neville filters [79], finer directional wavelet filters [99], DFB [33], contourlets [100], curvelets [101], dual tree-complex wavelet transform [102], rotated complex wavelet filters [103], wedgelets [104], ridgelets [105], bandlets [106] etc. However, most of the approaches have high computational complexity with redundancy.

1.8 Motivation

It is observed that iris consists of non-uniform spectral information due to its irregular and random characteristics (tiny crypts, freckles, radial furrows, radial streaks, collarette, pigment spots etc.). It is remarkable that most state-of-the-art iris recognition algorithms fail to perform in a non-cooperative environment, where the probability of acquiring non-ideal iris images is very high [8]. Thus, the main task of an iris recognition system is to provide effective and efficient representation of an iris image with less number of FRs. The extracted features should have high discriminating capability. Multi-resolution analysis (MRA) based technique is well suited to represent these iris structures. It is well known that DWT is a very powerful tool in MRA. The power of DWT is to offer high temporal localization for high frequencies and good frequency resolution for low frequencies. Most of the iris feature extraction techniques in the literature used Gabor and/or off-the-shelf wavelet bases to extract the iris features. *Although there is defined standard for raw iris data, but there is none regarding iris feature representation* [21]. Thus, many issues are still open in the field of iris feature-extraction and the choice of filter bank (FB). The design of FBs and investigations of their properties (near-orthogonality, regularity, time-frequency localization, linear phase etc.) for image-coding, denoising, compression etc. have been carried out by many researchers. However, effectiveness of these properties in iris pattern recognition has not been addressed in the literature. Several non-ideal factors associated with iris images may increase the FRR. Hence, this book focused on the design of two-channel separable and non-separable FBs for the effective and efficient iris representation and *k-out-of-n*: A post-classifier to reduce the FRR.

1.9 Summary

This chapter presented the review of iris recognition algorithms, two-channel separable and non-separable filter banks.

References

1. Jain A, Ross A, Prabhakar S (2004) An introduction to biometric recognition. IEEE Trans Circuits Syst Video Technol 14(1):4–20
2. Clarke R (1994) Human identification in information systems: management challenges and public policy issues. Inf Technol People 7(4):6–37
3. Jain A, Bolle R, Pankanti S (1999) Personal identification in networked society. Kluwer Academic Publisher, Boston
4. Maltoni D, Maio D, Jain A, Prabhakar S (2005) Handbook of fingerprint recognition. Springer, New York
5. Daugman JG (1993) High confidence visual recognition of persons by a test of statistical independence. IEEE Trans Pattern Anal Mach Intell 25(11):1148–1161
6. Daugman JG (2004) How iris recognition works. IEEE Trans Circuits Syst Video Technol 14(1):21–30
7. Daugman JG (2003) The imporance of being random: statistical principles of iris recognition. Pattern Recogn 36(2):279–291
8. Roy K (2011) Recognition of non-ideal iris images using shape guided approach and game theory. Ph.D. dissertation, Department of Computer Science and Software Engineering, Concordia University, Canada
9. Proenca H (2006) Towards non-cooperative biometric iris recognition. Ph.D. dissertation, Department of Computer Science, University of Beira Interior, Portugal
10. Wildes RP (1997) Iris recognition: an emerging biometric technology. Proc IEEE 85(9): 1348–1363
11. Boles WW (1998) A human identification technique using images of the iris and wavelet transform. IEEE Trans Sig Process 46(4):1185–1188
12. Mallat S (1989) A theory for multiresolution signal decomposition: the wavelet representation. IEEE Trans Pattern Anal Mach Intell 11(7):674–694
13. Sanchez-Avila C, Sanchez-Reillo R, de Martin-Roche D (2001) Iris recognition for biometric identification using dyadic wavelet transform zero-crossing. In: Proceedings of the 35th IEEE international Carnahan conference on security technology, pp 272–277
14. Masek L (2003) Recognition of human iris patterns for biometric identification. Master's thesis, Department of Computer Science and Software Engineering, The University of Western Australia
15. Ma L, Tan T, Wang Y, Zhang D (2004) Personal identification based on iris texture analysis. IEEE Trans Pattern Anal Mach Intell 25(12):2519–2533
16. Ma L, Wang Y, Zhang D (2004) Efficient iris recognition by characterizing key local variations. IEEE Trans Image Process 13(6):739–750
17. Ma L, Tan T, Zhang D, Wang Y (2004) Local intensity variation analysis for iris recognition. Pattern Recogn 37(6):1287–1298
18. Ma L, Wang Y, Tan T (2002) Iris recognition based on multichannel gabor filtering. In: Proceedings of Asian conference on computer vision. Melbourne, Australia, pp 279–283
19. Ma L, Wang Y, Tan T (2002) Iris recognition using circular symmetric filters. In: Proceedings of the 25th international conference on pattern recognition (ICPR02), pp 414–417
20. Sun Z, Wang Y, Tan T, Cui J (2005) Improving iris recognition accuracy via cascaded classifiers. IEEE Trans Syst Man Cybern B Cybern 35(3):435–441

21. Sun Z, Tan T (2009) Ordinal measures for iris recognition. IEEE Trans Pattern Anal Mach Intell 31(12):2211–2226
22. Dong W, Tan T, Sun Z (2010) Iris matching based on personalized weight map. IEEE Trans Pattern Anal Mach Intell 99(1):1–14
23. He Z, Sun Z, Tan T, Qiu X, Zhang C (2008) Boosting ordinal features for accurate and fast iris recognition. In: Proceedings of IEEE conference on computer vision and pattern recognition, pp 1–8
24. Vatsa M, Singh R, Noore A (August 2008) Improving iris recognition performance using segmentation, quality enhancement, match score fusion, and indexing. IEEE Trans Syst Man Cybern B 38(4):1021–1035
25. Schuckers S, Schmid N, Abhyankar A, Dorairaj V, Boyce C, Hornak L (2007) On techniques for off angle compnesation in nonideal iris recognition. IEEE Trans Syst Man Cybern B Cybern 37(5):1176–1190
26. Abhyankar A, Schuckers S (2007) A novel biorthogonal wavelet network system for off-angle iris recognition. Pattern Recogn 43:586–594
27. Velisavljević V (2009) Low-complexity iris coding and recognition based on directionlets. IEEE Trans Inf Forensics Secur 4(3):410–417
28. Velisavljević V, Beferull-Lozano B, Vetterli M, Dragotti P (2006) Directionlets: anisotropic multidirectional representation with separable filtering. IEEE Trans Image Process 15(7):1916–1933
29. Monro DM, Rakshit S (2007) DCT based iris recognition. IEEE Trans Pattern Anal Mach Intell 29(4):586–595
30. Proença H, Alexandre LA (2007) Toward non-cooperative iris recognition: a classification approach using multiple signatures. IEEE Trans Pattern Anal Mach Intell 29(4):607–612
31. Huang J, You X, Yuan Y, Yang F, Lin L (2010) Rotation invariant iris feature extraction using guassian markov random fileds with non-seprable wavelet. Neurocomputing 73:883–894
32. Park C, Lee J, Oh S, Song Y (2003) Iris feature extraction and matching based on maltiscale and directional image representation. In: Proceedings of fourth international conference on scale space methods in computer vision. Springer, Isle of Skye, UK, pp 576–583
33. Bamberger R, Smith M (1992) A filter bank for the directional decomposition of images: theory and design. IEEE Trans Signal Process 40:882–893
34. Nabti M, Ghouti L, Bouridane A (2008) An effective and fast iris recognition system based on a combined multiscale feature extraction technique. Pattern Recogn 41:868–879
35. Thornton J, Savvides M (2007) A bayesian approach to deformed pattern matching of iris images. IEEE Trans Pattern Anal Mach Intell 29(4):596–606
36. Tisse C, Martin L, Torres L, Robert M (2002) Person identification technique using human iris recognition. In: Proceedings of the 25th international conference on vision interface, pp. 294–299
37. Lim S, Lee K, Byeon O, Kim T (2001) Efficient iris recognition through improvement of feture vector and classifier. ETRI 23(2):61–70
38. Helen S, Selvan S (2006) Iris feature extraction based on directional image representation. GVIP J 8(4):55–62
39. Altan A (2008) Recognition of selected fingerprints and iris features enhanced by curvelet transform with artificial neural network. In: Proceedings of 15th international conference on systems, signals, and image processing, pp. 421–424
40. Chen Y, Dass S, Jain A (2006) Localized iris image quality using 2-D wavelets. In: Proceedings of international conference on biometrics, pp 373–381
41. Du Y, Ives R, Etter D, Welch T (2006) Use of one-dimensional iris signatures to rank iris pattern similarities. Opt Eng 45(3):037 201-1-037 201–11
42. Wang F, Yao X, Han J (2007) Minimax probability machine multialgorithmic fusion for iris recognition. Inf Technol J 6(7):1043–1049
43. Ariyapreechakul P (2007) An improvement of iris pattern identification using radon transform. ECTI Trans Comput Inf Technol 3(1):45–50

44. Sung H, Lim J, Park J, Lee Y (2004) Iris recognition using collarette boundary localization. In: Proceedings of international conference on pattern recognition
45. Kumar A, Passi A (2010) Comparison and combination of iris matchers for reliable personal authentication. Pattern Recogn 43(3):1016–1026
46. Poursaberi A, Araabi B (2007) Iris recognition for partially occluded images: methodology and sensitivity analysis. EURASIP J Adv Sig Process 2007:1–12
47. Dey S, Samanta D (2010) Improved feature processing for iris biometric authentication system. World Acad Sci Eng Res 4:486–493
48. Bodade R, Talbar S (2009) Iris recognition using combination of dual tree rotated complex wavelet and dual tree complex wavelet. In: Proceedings of the IEEE international conference on communications, pp 5425–5429
49. Bae K, Noh S, Kim J (2003) Iris feature extraction using independent component analysis. In: Proceedings of international conference on audio and video based biometric person authentication, pp 838–844
50. Dorairaj V, Schmid N, Fahmy G (2005) Performance evaluation of iris based recognition system implementing global ICA encoding. In: Proceedings of international conference on image processing, vol. 3, pp 285–288
51. Krichen E, Garcia-Salicetti S, Dorizzi B (2009) A new phase-correlation-based iris matching for degraded images. IEEE Trans Syst Man Cybern B Cybern 39(4):924–934
52. Chou C, Shih S, Chen W, Cheng V, Chen D (March 2010) Non-orthgonal view iris recognition system. IEEE Trans Circuits Syst Video Technol 20(3):417–430
53. Bowyer K, Hollingsworth K, Flynn P (2008) Image understanding for iris biometrics: a survey. Comput Vis Image Underst 110(2):281–307
54. Brune O (1931) Synthesis of a finite two terminal network whose driving point impedence is a prescribed function of frequency. J Math Phys 10:191–235
55. Swamy M, Rajan P (1986) Symmetry in two-dimensional filters and its application. Mercel Dekker Inc., New York
56. Croisier A, Esteban D, Galand C (1976) Perfect channel splitting by use of interpolation, decimation, tree decomposition techniques. In: Proceedings of IEEE conference on information systems, pp 443–446
57. Crochiere R, Rabiner L (1983) Multirute digital signal processing. Prentice-Hall, Englewood Cliffs
58. Vetterli M (1984) Multidimensional subband coding: some theory and algorithms. Signal Process 6:97–112
59. Ramstad T (1984) Analysis/synthesis filter banks with critical sampling. In: IEEE conference on digital signal processing. Florence, pp 130–134
60. Smith M, Barnwell T (1985) A unifying framework for analysis synthesis systems based on maximally decimated filter banks. In: Proceedings of IEEE conference on IEEE ICASSP, vol 10, pp 521–524
61. Smith M, Barnwell T (1987) A new filter bank theory for time frequency representation. IEEE Trans Acoust Speech Signal Process 35(3):314–327
62. Vetterli M (1985) Splitting a signal into subsampled channels allowing perfect reconstruction. In: Proceedings of the SPIE conference on signal processing digital filtering, Paris
63. Nayebi K, Barnwell T, Smith M (1992) Time-domain filter-bank analysis: a new design theory. IEEE Trans Signal Process 40:1412–1429
64. Vaidyanathan PP (1987) Quadrature mirror filter-banks, m-band extensions, and perfect-reconstruction techniques. IEEE ASSP Mag 4:4–20
65. Vetterli M, Kovacevic J (1995) Wavelets and subband coding. Prentice-Hall, Englewood Cliffs
66. Nguyen T, Vaidyanathan P (1989) Two channel pr FIR QMF structures which yield linear phase analysis and synthesis filters. IEEE Trans Acoust Speech Signal Process 37:676–690
67. Patil B, Patwardhan P, Gadre V (2008) On the design of fir wavelet filter banks using factorization of a halfband polynomial. IEEE Signal Process Lett 15:485–488
68. Daubechies I (1992) Ten lectures on wavelets. SIAM, Philadelphia

69. Daubechies I, Feauveau J (1992) Biorthogonal bases of compactly supported wavelets. Commun Pure Appl Math 45:485–560
70. Ansari R (1991) Wavelet construction using lagrange halfband filters. IEEE Trans Circuits Syst Express Briefs 38:1116–1118
71. Sweldens W (1996) The lifting scheme: a custom-design construction of biorthogonal wavelet. J Appl Comput Harmon Anal 3(2):186–200
72. Daubechies I, Sweldens W (1996) Factoring wavelet transforms into lifting steps. J Fourier Anal Appl 4:1–24
73. Phoong S, Kim C, Vaidyanathan P, Ansari R (1995) A new class of two-channel biorthogonal filter banks and wavelet bases. IEEE Trans Signal Process 43(3):649–665
74. Ansari R, Kim C, Dedovic M (1999) Structure and design of two-channel filter banks derived from a triplet of halfband filters. IEEE Trans Circuits Syst II Analog Digit Signal Process 46(12):1487–1496
75. Tay DBH, Palaniswami M (2004) A novel approach to the design of the class of triplet halfband filterbanks. IEEE Trans Circuits Syst Syst II Express Briefs 51(7):378–383
76. Tay DBH (2008) ETHFB: a new class of even-length biorthogonal wavelet filters for hilbert pair design. IEEE Trans Circuits Syst Syst I Regul Pap 55(6):1580–1588
77. Chan S, Yeung K (2004) On the design and multiplierless realizaton of perfect reconstruction triplet-based fir filterbanks and wavelet bases. IEEE Trans Circuits Syst Syst I 51(8): 1476–1491
78. Kha H, Tuan H, Nguyen T (2011) Optimal design of fir triplet halfband filter bank and application in image coding. IEEE Trans Image Process 22(2):586–591
79. Eslami R, Radha H (2010) Design of regular wavelets using a three-step lifting scheme. IEEE Trans Signal Process 58(4):2088–2101
80. Kovacevic J, Sweldens W (2000) Wavelet families of increasing order in arbitrary dimensions. IEEE Trans Image Process 9(3):480–496
81. Nguyen T (1995) Digital filter banks design: quadratic constrained formulation. IEEE Trans Signal Process 43:2103–2108
82. Viscito E, Allebach J (1991) The analysis and design of multi-dimensional fir perfect reconstruction filter-banks for arbitrary sampling lattices. IEEE Trans Circuits Syst 38:29–41
83. Vaidyanathan PP (1993) Multirate systems and filter banks. Prentice-Hall, Englewood Cliffs
84. Soman A, Vaidyanathan P (1993) Linear phase paraunitary filter banks: theory, factorizations and designs. IEEE Trans Signal Process 41(12):3480–3496
85. Karlsson G, Vetterli M (1990) Theory of two-dimensional multirate filter banks. IEEE Trans Acoust Speech Signal Process 38(6):925–937
86. Kovacevic J, Vetterli M (1995) Nonseparable two and three dimensional wavelets. IEEE Trans Signal Process 43(52):1269–1272
87. Hong P, Smith T (2002) An octave band family of non-redundant directional filter banks. In: Proceedings of international conference on acoustics, Speech and signal processing, vol 2, pp 1165–1168
88. Nguyen T, Oraintara S (2005) Multiresolution direction filterbanks: theory, design and applications. IEEE Trans Signal Process 53(10):3895–3905
89. Kalker T, Shah I (1996) A group theoretic approach to multidimensional filter banks: theory and applications. IEEE Trans Signal Process 44(6):1392–1405
90. Horn R, Johnson C (1999) Matrix analysis. Cambridge University Press, Cambridge
91. Zhou J, Do MN, Kovacevic J (2006) Special paraunitary matrices, Cayley transform, and multidimensional orthogonal filter banks. IEEE Trans Image Process 15(2):511–519
92. Zhou J, Do MN, Kovacevic J (2005) Multidimensional orthogonal filter-bank characterization and design using the cayley transform. IEEE Trans Image Process 14(6):760–769
93. Lin Y, Vaidyanathan P (1996) Theory and design of two-parallelogram filter banks. IEEE Trans Signal Process 44(11):2688–2705
94. Ikehara M (1995) Modulated two-dimensional perfect reconstruction FIR filter banks with permissible passbands. In: Proceedings of international conference on acoustics, speech, and signal processing, pp 1468–1471

95. Shah I, Kalker T (1993) Theory and design of multidimensional QMF sub-band filters from 1D filters and polynomials using transforms. IEE Proc Commun Speech Vision 140(1):67–71

96. Tay DBH, Kingsbury N (1993) Flexible design of multidimensional perfect reconstruction FIR 2-band filter-banks using transformation of variables. IEEE Trans Image Process 2:466–480

97. Mersereau R, Mecklenbrauker W, Quatieri T (1976) Mcclellan transformation for 2d filtering: i-design. IEEE Trans Circuits Syst 23(7):405–414

98. Eslami R (2007) A new family of nonredundant transforms using hybrid wavelets and directional filter banks. IEEE Trans Image Process 16(4):1152–1167

99. Lu Y, Do M (2005) The finer directional wavelet transform. In: Proceedings of the IEEE ICASSP, Philadelphia

100. Do M, Vetterli M (2005) The contourlet transform: an efficient directional multiresolution image representation. IEEE Trans Image Process 14(12):2091–2106

101. Candes E, Demanet L, Donoho D (2005) Fast discrete curvelet transform. Technical report, CalTech

102. Kingsbury N (2002) Complex wavelets for shift invariant analysis and filtering of signals. J Appl Comput Harmon Anal 10(3):234–253

103. Kokare M, Biswas P, Chatterji B (2005) Texture image retrieval using new rotated complex wavelet fitlers. IEEE Trans Syst Man Cybern B Cybern 35(6):1168–1178

104. Donoho DL (1999) Wedgelets: nearly minimax estimation of edges. Ann Stat 27(3):859–897

105. Candes E, Donoho DL (1999) Ridgelets: a key to higher dimensional intermittency? Philos Trans R Soc Lond 357:2495–2509

106. Pennec E, Mallat S (2005) Sparse geometric image representation with bandelets. IEEE Trans Image Process 14(4):423–438

Chapter 2
Features Based on Triplet Half-Band Wavelet Filter-Banks

Abstract This chapter presents the design of a new class of triplet half-band filter bank (THFB) and investigates its properties to extract iris image features. The feature extraction process and post-classifier have been discussed in this chapter.

Keywords Iris recognition · *k-out-of-n:A* classifier · Post-classifier · THFB · Wavelets

2.1 Introduction

This chapter presents the design of a new class of triplet half-band filter bank (THFB) and investigates its properties to extract the iris image features. The feature extraction process and the post-classifier have been discussed in this chapter.

Daugman [1] used multi-scale quadrature 2-D Gabor filter to demodulate phase information of an iris image to create IrisCode for the authentication. Proenca and Alexandre [2] partitioned normalized iris image into six regions and obtained six IrisCodes using Gabor filters. The matching scores of these six regions are fused together to generate an overall matching score. However, Gabor basis provides an over-complete representation which increases the redundancy and thus time required for iris feature extraction is high. Masek [3] introduced the Log-Gabor filter to encode the phase information of an iris image. Vatsa et al. [4] derived two types of distinct iris features (Log-Gabor and Euler numbers) from the normalized iris images and improved the recognition accuracy by effective segmentation technique, quality enhancement, and SVM rule. However, it consists of many stages to improve the performance. Boles and Boashash [5] used 1-D WT to compute the zero-crossing representation at different resolution levels of a concentric circle on an iris image. However, this method provides very less information along a virtual circle on the iris which affects the recognition accuracy. Wildes [6] obtained the characterization of iris texture through Laplacian pyramid with four different resolution levels. Furthermore, image registration technique is used to align the feature vectors of two iris images and compared the corresponding feature vectors using a normalized

A. D. Rahulkar and R. S. Holambe, *Iris Image Recognition*,
SpringerBriefs in Signal Processing, DOI: 10.1007/978-3-319-06767-4_2,
© The Author(s) 2014

correlation. However, this registration technique significantly increases the computational complexity of the entire method. Lim et al. [7] used 2-D Haar WT to decompose an iris image into four levels and encoded the fourth-level high-frequency information into an 87-bits binary ordinal code depending upon the sign of the filtered results. However, this method looses middle frequency components of the iris. Ma et al. [8] extracted the texture features of an iris using a bank of spatial filters and used quality descriptor, bootstrap learning and FLD to improve the recognition rate. However, this method is not suitable to perform well in the presence of eyelids/eyelashes occlusion. The same authors used 1-D quadratic spline WT along the angular direction of a normalized iris image [9]. In their work, feature vectors derived based on the local sharp variation points of a variable length called as shape code (SC). Nabti et al. [10] presented multi-resolution iris feature extraction technique by analyzing the iris using first wavelet maxima components and then applying a special Gabor filter bank on the normalized iris image to extract all dominant features. Velisavljević [11] presented iris coding and recognition using directionlets based on 9/7 bi-orthogonal wavelet basis. Abhyankar and Schuckers [12] introduced biorthogonal wavelet neural network (BWN) for off-angle iris recognition by adjusting non-ideal factors through repositioning the BWN. Monro et al. [13] presented ordinal encoding scheme based on the difference of optimized DCT coefficients of overlapped angular patches from normalized iris image. Sun and Tan [14] presented OMs for iris feature representation scheme based on MLDFs using 2-D Gaussian filters. MLDF has been used on 1,024 densely sampled image regions to obtain 1024 bits ordinal code for every iris image with flexible interlobe distance (d). This method has achieved the good trade-off between distinctiveness and robustness. However, this representation may lose some image specific information. Dong et al. [15] introduced a personalized iris matching strategy using a class-specific weight map learned from the training images of the same iris class. The robustness of the weight map totally depends upon the number of training images within a class. However, this method requires more number of training images to decide an effective weight-map within a class which leads to increase the computational cost. Kumar and Passi [16] presented a comparative study of the iris recognition performance using Log-Gabor, Haar wavelet, DCT, and FFT based phase ordinal encoding with small number of training images. Huang et al. [17] introduced a rotation invariant approach for iris feature extraction scheme based on non-separable wavelet FB and Gaussian Markov random field (GMRF). In their work, non-separable wavelet is constructed by using centrally symmetric orthogonal matrices to obtain the iris features in eight different directions. Furthermore, FLD with polynomial kernel is used to improve the computational efficiency and classification accuracy. Park et al. [18] noticed the importance of capturing directional information in iris images where a DFB is applied to a band-pass filtered iris images to derive feature vectors.

It is observed that most of these iris recognition algorithms are sensitive to iris segmentation (detection of inner and outer boundaries). The noise due to inaccurate detection of outer boundary can be easily removed, but the inaccurate detection of pupillary boundary plays a very important role for the iris verification. Due to the inaccurate detection of pupillary boundary, either iris information will lose or

Fig. 2.1 Block diagram of
two-channel PRFB

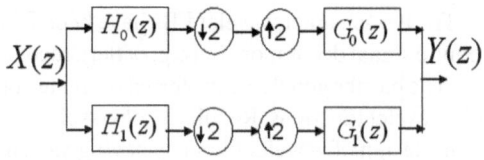

occlusion of pupil will occur on the normalized iris. Thus, an accurate segmentation of iris region is significantly important for iris recognition which may not be possible in the practical iris recognition system. The great progress has been made on iris feature representation, but it is still an open problem. *Although, there is a defined standard for raw iris data, but there is none regarding iris features representation* [14]. Most of the iris feature extraction techniques in the literature used off-the-shelf wavelet basis to extract the iris features. However, many issues are still open in the field of iris feature-extraction and the choice of FB. The design of FB and investigations of their properties (near-orthogonality, regularity, time-frequency localization, linear phase, etc.) for image-coding, denoising, compression, etc. have been carried by many researchers. However, effectiveness of these properties in iris pattern recognition is not addressed in the literature. Hence, this chapter introduces a new wavelet basis for iris texture representation in an attempt to extract effective and compact iris features, and a flexible post-classifier *k-out-of-n:A* so as to handle possible artifacts especially segmentation error (inaccurate detection of inner and outer boundaries of iris), occlusion of eyelids/eyelashes, reflection on iris, shadow of upper/lower eyelid on iris, non-linear deformation, etc.

2.2 Review of the Related Filter Banks

The block diagram of two channel Filter Banks (FB) is shown in Fig. 2.1. The necessary and sufficient conditions for PR are given by the following two equations:

$$H_0(z)G_0(z) + H_0(-z)G_0(-z) = 2, \tag{2.1}$$

$$H_1 = z^{-1}G_0(-z), \quad G_1(z) = zH_0(-z). \tag{2.2}$$

The product filter $P(z) = H_0(z)G_0(z)$ in Eq. (2.1) belongs to the special class of filters known as half-band filter. The analysis scaling and wavelet functions are given by the following two-scale dilation equations:

$$\phi(t) = \frac{2}{|H_0(\omega)|_{\omega=0}} \sum_n h_0(n)\phi(2t - n), \tag{2.3}$$

$$\psi(t) = \frac{2}{|G_0(\omega)|_{\omega=0}} \sum_n h_1(n)\phi(2t - n). \tag{2.4}$$

where $h_0(n)$ and $h_1(n)$ are the analysis low-pass filter (LPF) and high-pass filter (HPF) coefficients respectively.

The two-channel wavelet FBs have been widely used in many applications which require signal decomposition e.g. pattern recognition, compression, super-resolution, etc. The biorthogonal FBs preferred for image processing due to its symmetric scaling and wavelet functions. Regularity of wavelets is an important property which imposed in the design of wavelet filters by having vanishing moments (zeros at $z = -1$) at the aliasing frequencies of wavelet HPFs. Meanwhile, one of the design characteristics of wavelet FBs for signal analysis is the ability to achieve time-frequency localization. Thus, it is desirable to design a wavelet FBs that fulfill PR, linear phase, orthogonality, regularity and time-frequency localization. One of the most common design methods for designing wavelet filters is the factorization of a HBP [19]. It is well known that most of the popular biorthogonal FBs (e.g. Cohen-Daubechies-Feauveau (CDF-9/7), spline family of wavelet FBs [20–22] are designed by the factorization of a Lagrange half-band polynomial (LHBP). The LHBP has the maximum number of zeros at $z = -1$ to achieve the maximum regularity.

However, LHBP filters do not have any degree of freedom and thus there is no direct control over frequency response of the filters. In order to have some independent parameters (which can be optimized to obtain some control over frequency response of the filter), Patil et al. [19] used general half-band filter factorization (not LHBP) to design two-channel bi-orthogonal wavelet FIR FBs (BWFB). However, it is observed that this factorization (includes decision of factorization of remainder polynomial and reassignment of zeros) improves the frequency response of one of the filters (analysis/synthesis) at the cost of the other filter (synthesis/analysis). The improvement in frequency response of both the filters heavily depends upon the factorization of a HBP. This is somewhat tedious task for higher order HBPs. It is also observed that the designed biorthogonal FBs lose "maximal flat" condition and it is far away from orthogonality condition. Lifting scheme is also one of the attractive schemes to design wavelet FBs that provides structurally imposed PR property [23]. Based on the lifting scheme, Phoong et al. [24] introduced a class of halfband pair filterbank (HPFB) structure based on two kernels. In HPFB structure, magnitudes of the frequency responses of the analysis and synthesis low pass filters at $\omega = \pi/2$ are restricted to 0.5 and 1.0 respectively or vice-versa. This encounters certain restrictions for the control of its frequency response. In order to overcome this restriction [24], Ansari et al. [25] proposed a FB structure based on a class of THFB using three kernels that have structural PR, feature-rich structure and simple design. In their work, two methods are proposed to design two-channel 1-D biorthogonal FBs based on the triplet of half-band filters. The first method is based on Lagrange filter coefficients to achieve maximal flatness in the filters (due to maximum regularity) with slow frequency roll-off. The second method is employed Remez algorithm which results in equiripple filters with user-defined cut-off frequencies. It has sharp frequency roll-off but does not have the regularity condition. In their work, the shape parameter p is used to achieve a greater flexibility in frequency response of the filters. However, regularity order related to the number of zeros at $z = -1$ has not been specified. In order to bridge the gap between these two extremes, Tay and Palaniswami [26] introduced a novel approach to design a class of THFB. Parametric Bernstein polynomial is used to impose the required vanishing moments (VMs). This technique is based on iterative least square approach to determine the coefficients of Bernstein polynomial. However, Bernstein polynomial is suitable for nearly maximally flat frequency response rather than for ripple responses with sharp roll-off. In [27], Tay modified the work of [26] and suggested a new class of even-length biorthogonal wavelet filters for Hilbert pair design (ETHFB). Chan and Yeung [28] presented a design of THFB with regularity using semi-definite programming (SDP). However, the semi-infinite constraints are approximated by a large finite number of constraints that give an inefficient design for higher order filters. Kha et al. [29] proposed an efficient SDP method in order to design a class of THFB with optimal frequency selectivity for a given regularity condition. Eslami and Radha [30] generalized Ansari's method to a multidimensional FB design with any number of analysis and synthesis VMs using a structural approach based on Kovacevic method [31]. The filters designed in [30] achieved better regularity, lower frame bounds and better frequency selectivity than the filters designed in [31].

In this chapter, a new class of THFB is designed in order to overcome the limitations of recently presented FBs in [19, 25]. Firstly, three kernels are designed from the generalized HBP by imposing the zeros at $z = -1$. These three designed kernels are used in three step ladder structure to design a new class of THFB by varying the shape parameter p. The objective function involves optimization of frequency response of the filters. We have also shown that the designed filters achieve better frequency selectivity, near-orthogonality, good time-frequency localization with linear phase and PR condition. This FB is used to form a new wavelet basis for extracting the textural features of an iris image.

2.2.1 Triplet Halfband Filter Bank

The analysis and synthesis LPFs of a class of Triplet Halfband Filter Bank (THFB) consist of three kernels as follows [25] :

$$H_0(z) = \frac{1+p}{2} + \left(\frac{1+p}{2}\right) z T_1(z^2) \left(\frac{1 - pzT_0(z^2)}{1+p}\right), \tag{2.5}$$

$$G_0(z) = \frac{1 + pzT_0(z^2)}{1+p} + \frac{1-p}{1+p} z T_2(z^2) H_0(-z). \tag{2.6}$$

where, $T_m(z)$ is obtained using:

$$T_m(z) = \sum_{n=1}^{N \in \, even} t_m(n)(z^{-n} + z^{n-1}), \quad m = 0, 1, 2$$

The three HBPs $T_0(z^2)$, $T_1(z^2)$, and $T_2(z^2)$ required for Eqs. (2.5) and (2.6) are obtained using upsampled by 2 operation as:

$$zT_m(z^2) = \sum_{n=1}^{N_m \in \; even} t_m(n)(z^{-2n+1} + z^{2n-1}), \quad m = 0, 1, 2. \tag{2.7}$$

and the coefficients $t_m(n)$ are obtained using standard Lagrange interpolation formula as:

$$t_m(n) = \frac{(-1)^{n+N_m-1} \prod_{l=1}^{2N_m}(N_m + 1/2 - l)}{(N_m - n)!(N_m - 1 + n)!(2n - 1)}$$

The analysis and synthesis HPFs are obtained from Eq. (2.2) by quadrature mirroring the LPFs. This class of THFB is implemented using three-step ladder structure. The parameter p (degree of freedom) provides some flexibility in order to set the same magnitude of the frequency response at $\omega = \pi/2$ for both the analysis and synthesis LPFs (symmetry between analysis and synthesis filters).

2.2.2 Factorization Based on a Generalized Half-Band Polynomial

The alternate approach to the design of two-channel FBs is the design of a halfband filter, followed by its factorization to derive analysis and synthesis LPFs. Regularity imposed in the design of $P(z)$ by introducing zeros at $z = -1$. Patil et al. [19] presented an approach to design FIR wavelet FBs using factorization of a half-band polynomial (HBP). This approach is briefly discussed as follows:

1. Assume generalized symmetric half-band polynomial $P(z)$ of order K as

$$P(z) = a_0 + a_2 z^{-2} + \cdots + a_{\frac{K}{2}-1} z^{-\frac{K}{2}-1} + z^{-\frac{K}{2}} + a_{\frac{K}{2}-1} z^{-\frac{K}{2}+1} + \cdots + a_0 z^{-K}$$

   The coefficients a_k (degree of freedom) of $P(z)$ have to be designed.
2. Obtain L constraints on the coefficients a_k of $P(z)$ by introducing L zeros at $z = -1$.
3. Using these constraints, $P(z)$ can be expressed in terms of the independent (free) parameters.
4. Now express the polynomial $P(z)$ as $P(z) = (z + 1)^L R(z)$, where $R(z)$ is the remainder polynomial. Next factorize this $R(z)$ into $R_1(z)$ and $R_2(z)$ to obtain final filters as $H_0(z) = (z + 1)^{L/2} R_1(z)$ and $G_0(z) = (z + 1)^{L/2} R_2(z)$.

   The independent coefficients are obtained by optimizing the objective function (frequency responses of the factored filters $H_0(z)$ and $G_0(z)$). It is observed that the improvement in frequency responses of both the filters $H_0(z)$ and $G_0(z)$ is totally depend on the step 4.

2.3 Design of New Class of THFB

In this chapter, a new class of THFB is formulated in order to avoid the factorization process and improve the frequency response of both the filters simultaneously. First, general HBP of an order K (expressed in coefficients a_k) which offers $(\frac{K}{2}+1)$ degrees of freedom to impose the vanishing moments is considered. From this polynomial, three HBPs $P_1(z)$, $P_2(z)$, and $P_3(z)$ are obtained by imposing M zeros at $z = -1$, where $M < (\frac{K}{2}+1)$. With this, desired number of independent parameters a_k (degree of freedom) are obtained without imposing maximum flatness constraint. These three HBPs can be expressed as follows:

$$P_i(z) = (z^{-1} + 1)^{M_i} R_i(z), \quad i = 1, 2, 3. \tag{2.8}$$

where M_i is the number of zeros at $z = -1$ for ith polynomial and the remainder term $R_i(z)$ is given by the following equation:

$$R_i(z) = a_0 + c_1 z^{-1} + c_2 z^{-2} + \cdots + a_0 z^{K-M_i} \tag{2.9}$$

where c_j are the constants which can be expressed as functions of a single parameter a_0 (as given in Eq. (2.9)). Thus, three remainder polynomials $R_1(z)$, $R_2(z)$, and $R_3(z)$ are obtained by imposing M_1, M_2, and M_3 (where $M = M_1 + M_2 + M_3 = \frac{K}{2}$ can be a choice) zeros on $P(z)$ (of order K). It may be noted that the remainder polynomials can also be expressed into any desired number of independent or free parameters $(a_0, a_2, a_4, \dots)$. Expressing the c_j (remainder polynomial) with more number of a_k provides better flexibility at the cost of computational complexity. With this, these three HBPs provide one degree of freedom (independent parameters) by which flexibility in frequency responses can be achieved. We can also construct three kernels from three different HBPs. In this context, any desired values of M_1, M_2, and M_3 can be used and available degrees of freedom are utilized to tweak the frequency response. The required class of three kernels given in Eq. (2.7) is obtained by following equation

$$T_0(z^2) = z^{K/2} P_1(z) - 1;$$
$$T_1(z^2) = z^{K/2} P_2(z) - 1;$$
$$T_2(z^2) = z^{K/2} P_3(z) - 1.$$

These three designed kernels are then used in the three-step lifting scheme to obtain a new class of analysis and synthesis LPFs respectively as follows:

$$H_0(z) = \frac{1+p}{2} + \frac{1+p}{2} \left(z^{K/2} P_2(z) - 1 \right) \left(\frac{1}{1+p} \left(1 + p \left(1 - z^{K/2} P_1(z) \right) \right) \right), \tag{2.10}$$

$$G_0(z) = \frac{1 - p + pz^{K/2}P_1(z)}{1 + p} + \left(\frac{1 - p}{1 + p}\right)\left(z^{K/2}P_3(z) - 1\right)H_0(-z). \quad (2.11)$$

where, K is the order of HBP. The use of a class of THFB provides one more degree of freedom (p) by which we can shape better frequency responses of the final filters. Thus, the suggested design offers more flexibility in the design of filters using two degrees of freedom (a_k and p). This method has been illustrated in Sect. 2.3.1 with one example. Due to the use of THFB and general HBP, frequency responses of both the filters have been improved simultaneously. These filters satisfy regularity, near-orthogonality, linear-phase, and PR properties. The resulting lengths of the analysis LPF $H_0(z)$ is $N_1 + N_2 - 1$ and synthesis LPFs $H_0(z)$ is $N_1 + N_2 + N_3 - 2$, where N_1, N_2, and N_3 are the lengths of $P_1(z)$, $P_2(z)$, and $P_3(z)$ respectively. It is observed that synthesis LPF $G_0(z)$ is long and smooth as compared to synthesis HPF $G_1(z)$. This setting is more desirable to avoid blocking, checkerboard, and ringing artifacts during signal reconstruction in lossy coding [32].

2.3.1 Design Example

In the following example, 6th order HBP is used to design the required kernels $T_0(z^2)$, $T_1(z^2)$, and $T_2(z^2)$.

Example: Consider $P(z)$ of order 6

$$P(z) = a_0 + a_2z^{-2} + z^{-3} + a_2z^{-4} + a_0z^{-6}. \quad (2.12)$$

This $P(z)$ is used to construct $P_1(z)$, $P_2(z)$, and $P_3(z)$ by imposing the zeros at $z = -1$ with the help of synthetic-division such that these three polynomials can be expressed in terms of a single independent parameter a_0. Consider $M_1 = 0$, $M_2 = 1$, and $M_3 = 2$ such that $M_1 + M_2 + M_3 = M = 3$, where $M < (\frac{K}{2} + 1)$.

The polynomials are expressed as follows:

$$P_1(z) = a_0 + (a_0 + 0.5)z^{-2} + (-a_0 + \frac{1}{2})z^{-4} + a_0z^{-6},$$

$$P_2(z) = (1 + z^{-1})(a_0 + (-a_0)z^{-1} + \frac{1}{2}z^{-2} + \frac{1}{2}z^{-3} + (-a_0)z^{-4} + a_0^{-5}), \quad (2.13)$$

$$P_3(z) = (1 + z^{-1})^2(a_0 + (-2a_0)z^{-1} + (2a_0 + \frac{1}{2})z^{-2} + (-2a_0)z^{-3} + a_0z^{-4}).$$

Now optimized value of $a_0 = -0.062499$ is obtained using MATLAB optimization routine *fminunc*. This value of a_0 minimizes the energy in the ripples of these kernels. The proposed transfer functions for the class of THFB are obtained by using Eqs. (2.5) and (2.6) to derive the LP analysis filter $H_0(z)$ and synthesis filter $G_0(z)$. The optimized value of shaping parameter $p = \sqrt{2} - 1$ is used to obtain the same

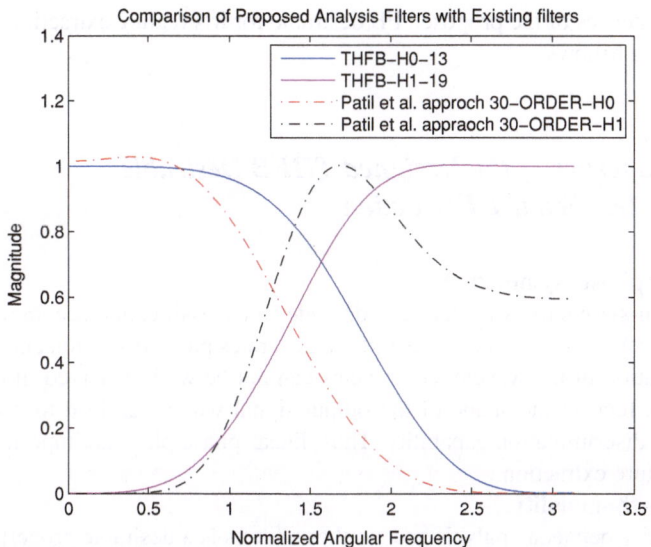

Fig. 2.2 Magnitude responses of the filter bank pair

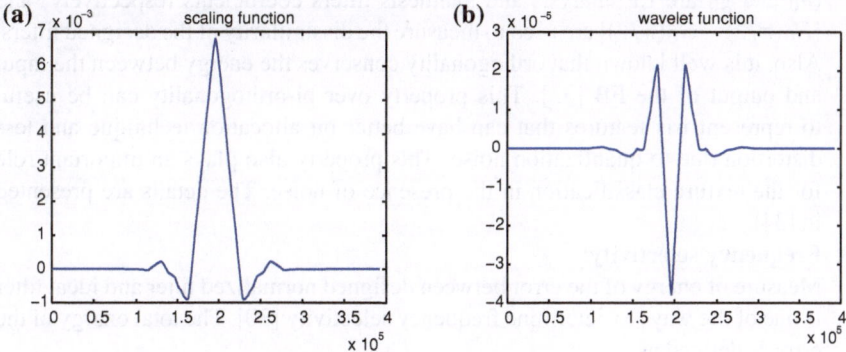

Fig. 2.3 a Scaling and **b** Wavelet functions

magnitude of $H_0(z)$ and $G_0(z)$ at $\omega = \pi/2$. The lengths of filters $H_0(z)$ and $G_0(z)$ are 13 and 19 respectively. We compare the results of this proposed FBs with the same length FBs designed using existing approach of [19]. The technique in [19] needs 30th order HBP to get analysis and synthesis LPFs of length 13 and 19 respectively. Fourteen zeros are imposed at $z = -1$ on 30th order HBP. Then six zeros to analysis and eight zeros to synthesis LPFs are reassigned to obtain 13/19 length filters respectively. The frequency responses of the proposed analysis LPF and HPF are compared with the existing FBs [19] as shown in Fig. 2.2. The analysis scaling and wavelet functions for the proposed FBs are shown in Fig. 2.3.

The properties of the proposed FB desirable for iris feature extraction are briefly discussed as follows:

2.3.2 Properties of the Designed THFB Desirable for Iris Feature Extraction

1. **Linear phase (symmetry)**:
   The non-symmetric wavelet basis degrades the classification accuracy due to its non-linear phase. The non-linear phase generates phase distortion and the spatial localization of the wavelet coefficients can not be well-preserved. It can have a major effect on the shape of the output signal which can lead to decrease the texture discrimination capability. Thus, linear-phase plays an important role for iris feature extraction.

2. **Near-orthogonality**:
   Similarity between analysis and synthesis filters is a desirable property. It could be used as a measure of near-orthogonality [30]. It is a quantitative measure of how far a biorthogonal FBs from orthogonality. In this chapter, $\|h_0 - g_0\|^2$ (h_0 and g_0 are LP analysis and synthesis filters coefficients respectively) and $|H_0(\pi/2) - G_0(\pi/2)|$ are used to measure the dissimilarity of the designed filters. Also, it is well known that orthogonality conserves the energy between the input and output of the FB [33]. This property over bi-orthogonality can be useful to represent iris features that can have better bit allocation technique and less distortion due to quantization noise. This property also plays an important role for the texture classification in the presence of noise. The details are presented in [34].

3. **Frequency selectivity**:
   Measure of energy of the error between designed normalized filter and ideal filter is one of the ways to determine frequency selectivity [30]. The total energy of the error is defined as

$$E = \int_0^{(\pi/2)} |1 - H_0(\omega)|^2 d\omega + \int_{\pi/2}^{\pi} |H_0(\omega)|^2 d\omega. \tag{2.14}$$

   In this chapter, Eq. (2.14) is used as an objective function to choose the value of a_0. The proposed FB provides good frequency selectivity which is helpful to represent effective iris features.

4. **Regularity**:
   It is well known that one of the important properties of a wavelet FB is regularity. The regularity leads to smooth scaling and wavelet functions. Regularity is imposed in the design of wavelet FB by imposing zeros at $z = -1$ [35]. The LP filtering followed by decimation will result in the aliasing due to the lower

Table 2.1 Properties measures of 1-D analysis LPFs

Properties measures	Patil's FB 30th order	Proposed THFB 13/19
$|H_0(pi/2) - G_0(pi/2)|$	0.6318	0.0026
$\|h_0 - g_0\|^2$	57.76	0.1589
E	61.65	59.65
Δt^2	0.5595	0.9464
$\Delta \omega^2$	0.5662	0.5672

sampling rate. Consequently, successively LP filtering with decimation results in more aliasing terms [34]. This leads to produce significant iris verification error. Thus, it is desirable to approximate the iris features in order to minimize the quantization error and have better iris texture discrimination. It is also important to note that if regularity order of analysis wavelet function is greater than the synthesis wavelet function, then the resultant wavelet basis has more approximation power in the decomposition section and is more regular in reconstruction [30]. This setting is more desirable for iris feature extraction.

5. **Time-Frequency localization**:
   Time-frequency localization plays a very important role in order to show the ability of WT for signal analysis. However, many of wavelet design techniques do not explicitly incorporate any localization criteria [36]. The measure of spatial localization and frequency localization are computed directly from filter coefficients. The details are given in [36, 37]. It is also given that lower $\Delta \omega^2$ have a sharper roll-off in the frequency response. The proposed FB provides better time-frequency resolution, so it can be well adopted to characterize the variations in iris images.

Table 2.1 presents the performance measures for some of the properties of the designed filters using approach in [19] and the class of THFB. It is observed that investigated filters provide more similarity between analysis and synthesis LPFs (near-orthogonality), good frequency selectivity, and good time-frequency localization. It may be noted that the designed filters can achieve all the above properties for any order of HBP.

In order to apply investigated class of THFB to iris images, 2-D extension of wavelets are required. An obvious way to construct separable 2-D wavelet filters is to use tensor product of their 1-D counterparts. A 2-D approximation and three detail functions are obtained from Eq. (2.15) as:

$$\begin{aligned} L(z) &= H_0^{1d}(z_1) \times H_0^{1d}(z_2) \\ H(z) &= H_0^{1d}(z_1) \times H_1^{1d}(z_2) \\ V(z) &= H_1^{1d}(z_1) \times H_0^{1d}(z_2) \\ D(z) &= H_1^{1d}(z_1) \times H_1^{1d}(z_2). \end{aligned} \tag{2.15}$$

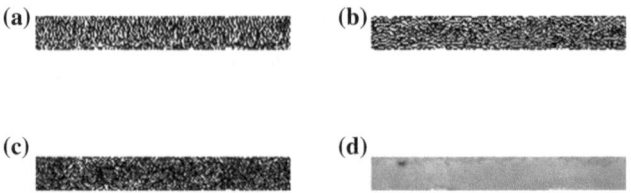

Fig. 2.4 One-level decomposition of CASIA-IrisV3.0 iris image using proposed class of THFB **a** LH sub-band **b** HL sub-band **c** HH sub-band **d** approximate (LL) sub-band

where $H_0^{1d}(z_1)$ and $H_1^{1d}(z_1)$ are 1-D LPF and HPF respectively of the designed filter-banks. The one-level decomposition results in *Vertical* (*V*), *Horizontal* (*H*), and *Diagonal* (*D*) sub-bands and one *approximation sub-band* (*L*), which corresponds to LH, HL, HH, and LL sub-bands respectively as shown in Fig. 2.4a–d for one of the iris images.

2.4 Iris Recognition Algorithm

In this work, iris features which are mostly oriented in vertical, horizontal and diagonal directions are computed by the introduced wavelet basis. The inner half iris region is divided into six sub-images and selected only four regions for further processing. This new class of THFB is applied separately on each of the four selected sub-blocks (sub-images). The feature vector for each sub-image is derived by estimating the channel energies of the THFB. The four distance scores are obtained and fused by the flexible post-classifier (*k-out-of-n:A*) in order to develop robust iris recognition technique. The block diagram of the half-iris recognition technique is shown in Fig. 2.5.

2.4.1 Feature Extraction Using a New Class of THFB

The original eye image must be preprocessed in order to extract iris features from an eye image. The preprocessing involves localization and normalization of iris image. In this work, iris is localized using Daugman's IDO and normalized with the help of DRSM of the fixed size [1]. The preprocessing steps are shown in Fig. 2.6a–c.

Although some of the existing methods extract iris texture efficiently, their performance degrades significantly when the image quality is poor. Chen et al. [38] suggested that different regions of the iris have different qualities and local iris image regions with better quality have better classification capability and vice-versa. In multi-biometric recognition system, fusion of information extracted from classifiers provide better recognition performance as compared to single classifier

Enrollment

Iris **Segmentation** and **Normalization**

↓

Selected **upper half** normalized iris image

↓

Partitioned half iris into six regions and selected only **four regions**

↓

Decomposed each sub-region using proposed THFB upto **two levels**

↓

Construct FV by computing **energies of subbands** for each sub-region

↓

Obtained four FVs of each length 3*no of levels+1

↓

Stored in **Database**

Testing

Iris **Segmentation** and **Normalization**

↓

Selected **upper half** normalized iris image

↓

Partitioned half iris into six regions and selected only **four regions**

↓

Decomposed each sub-region using proposed THFB upto **two levels**

↓

Construct FV by computing **energies of subbands** for each sub-region

↓

Obtained four FVs of each length 3*no of levels+1

↓

Obtained four **Canbera Distances** (CDs) and find minimum CD for each region.

↓

Verify the person using *k-out-n:A* postclassifier

Fig. 2.5 Block diagram of the proposed iris recognition scheme

Fig. 2.6 a Original eye image.
b Segmented iris image.
c Normalized iris image.
d Partitioned normalized iris image

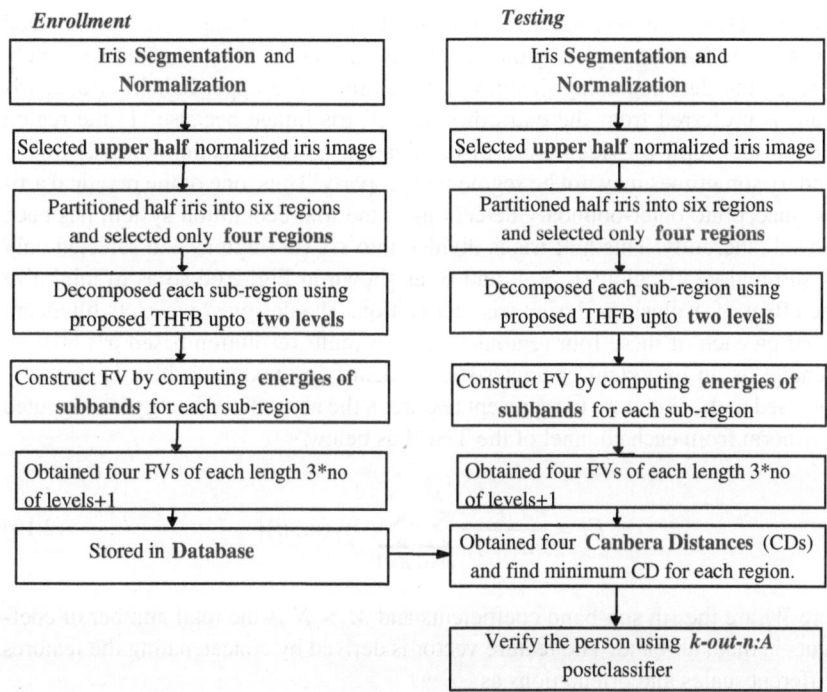

(a)   (b)

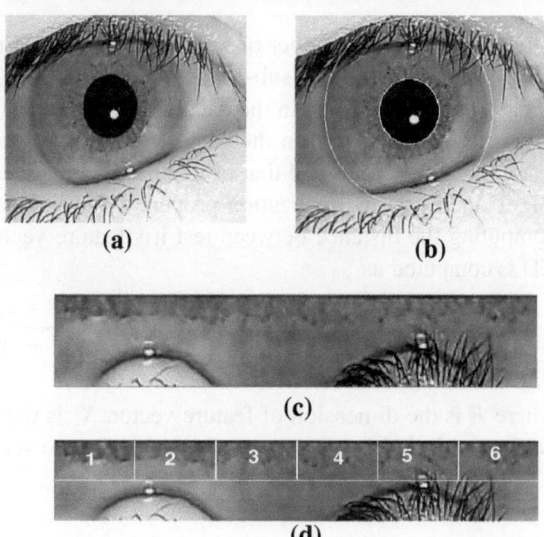

(c)

(d)

[2, 4, 39]. Therefore, instead of recognizing the entire iris image, we have divided the iris image into multiple regions. Each iris sub-region is recognized separately and fused the decision using flexible post-classifier. The upper half-iris (inner iris region) is preferred from the entire normalized iris image because (1) the region closer to the pupil provides more discriminating iris information and (2) limbic boundary sometimes may not be segmented properly. Thus, one of the practical arti-facts (inaccurate outer-boundary detection) in the iris recognition system has been removed inherently. This half-iris is divided into six sub-regions and selected only four sub-regions (Region: 1, 3, 4, and 6) as shown in Fig. 2.6d so as to minimize some effect of occlusions during iris recognition. The designed wavelets filters are applied on each of these four regions to extract multi-resolution based iris texture. As energy is an important characteristic in identifying texture (which is normally being used in the literature to represent textures), the normalized energy is computed by L_1 norm from each channel of the THFB as below.

$$E_i = \frac{1}{M \times N} \sum_{m=1}^{M} \sum_{n=1}^{N} |W_i(m, n)| \qquad (2.16)$$

where W_i are the ith sub-band coefficients and $M \times N$ is the total number of coef-ficients in that sub-band. The feature vector is derived by concatenating the features at different scales and orientations as

$$E = [E_{1,1}, \ E_{1,2}, \ E_{1,3}, \ldots, E_{S,3}, \ E_a]; \qquad (2.17)$$

where S is the total number of scales and E_a is the energy of an approximate sub-band. The total number of sub-bands for THFB is $3S + 1$. The derived feature vectors of each region are stored in the database as reference (enrollment process). The test iris pattern is classified on the basis of minimum Canbera distance (CD) between test iris feature vector and that of feature vectors stored in the database. The use of CD is due to the normalization property of individual feature components before computing the distance between test iris feature vector and that of databases. The CD is computed as

$$CD(X, Y) = \sum_{i=1}^{B} \frac{|X_i - Y_i|}{|X_i| + |Y_i|} \qquad (2.18)$$

where B is the dimension of feature vector. X_i is the ith component of test feature vector and Y_i is ith component of enrolled feature vector.

2.4.2 Design of k-out-of-n:A Post-classifier for Iris Recognition

Iris recognition algorithms have succeeded in achieving a low FAR. However, reducing the FRR remains a major challenge. FRR needs to reduce to make iris recognition algorithm more robust. Many researchers have suggested that fusion of information extracted from classifiers provided better recognition performance as compared to single classifier [2, 4, 15, 39, 40]. In this work, fusion at the decision level is explored using k-out-of-n:A post-classifier. The value of k can be varied upto n and hence it is a flexible post-classifier. The designed post-classifier works on the ROC curve directly. ROC is the indirect representation of the distance scores between the test and enrolled feature vectors. ROC is obtained by varying threshold values of the distance scores. EER is a general optimal operating point that indicates threshold of the distance score. Multiple ROCs obtained from n-iris regions are fused by the post-classifier in order to improve the performance. The performance of the iris recognition system is assessed by measuring the errors made by rejecting genuine users (FRR), accepting impostor users for a given value of threshold (FAR), and computing computational complexity.

The test iris is accepted if at least any k out of the n-region(s) is (are) accepted (flexible k-out-of-n:A, where $k \leq n$). The details of general k-out-of-n system for the reliability analysis is given in [41]. In this rule, person is authenticated only when any k regions $k = 1, 2, 3$, and 4 out of n-regions ($n = 4$) passes the test of iris recognition. This system recognizes the person if any k distance scores (CDs) out of n-distance scores (CDs) are less than or equal to the corresponding thresholds.

FR can only occur by k-out-of-n:A post-classifier when n-regions iris tests produce FRs. Thus, genuine person is rejected when the combinations of k out of n-tests fail to recognize correctly. Based on this assumption, k-out-of-n:A post-classifier is framed for the different values of k as:

$$FR = \{CD_1^I > Th_1 \cap CD_2^I > Th_2 \ldots, \cap CD_n^I > Th_n\}, \quad k = 1.$$

$$FR = \{(CD_1^I > Th_1 \cup CD_2^I > Th_2) \cap (CD_1^I > Th_1 \cup CD_3^I > Th_3)$$

$$\ldots, \cap(\cdots \cup CD_n^I > Th_n)\}, \quad k = 2$$

$$FR = \{(CD_1^I > Th_1 \cup CD_2^I > Th_2 \cup CD_3^I > Th_3) \quad (2.19)$$

$$\cap (CD_1^I > Th_1 \cup CD_3^I > Th_3 \cup CD_4^I > Th_4)$$

$$\ldots, \cap(\cdots \cup CD_n^I > Th_n)\}, \quad k = 3$$

$$FR = \{CD_1^I > Th_1 \cup CD_2^I > Th_2 \cup CD_3^I > Th_3 \cup CD_4^I > Th_4\}, \quad k = 4$$

where superscript I denotes the intra-class comparisons. The final fused FRR for $k = 1$ is computed by counting the number of CD_i^I greater than Th_i denoted as $C(CD_i^I, Th_i)$

$$FRR = \prod_{i=1}^{n} \frac{C(CD_i^I > Th_i)}{N_i} \quad (2.20)$$

where N is the total number of intra-class comparisons. Similarly, final FRRs are obtained for the different values of $k = 2, 3, 4$.

The FA occurs only if any k out of n-regions passes the test (k-CDs less than or equal to the corresponding thresholds). It is expressed as:

$$FA = \{CD_1^E \leq Th_1 \cup CD_2^E \leq Th_2 \ldots, \cup CD_n^E \leq Th_n\}, \quad k = 1.$$
$$FA = \{(CD_1^E \leq Th_1 \cap CD_2^E \leq Th_2) \cup (CD_1^E \leq Th_1 \cap CD_3^E \leq Th_3)$$
$$\ldots, \cup(\cdots \cap CD_n^E \leq Th_n)\}, \quad k = 2$$
$$FA = \{(CD_1^E \leq Th_1 \cap CD_2^E \leq Th_2 \cap CD_3^E \leq Th_3) \tag{2.21}$$
$$\cup (CD_1^E \leq Th_1 \cap CD_3^E \leq Th_3 \cap CD_4^E \leq Th_4)$$
$$\ldots, \cup(\cdots \cap CD_n^E > Th_n)\}, \quad k = 3$$
$$FA = \{CD_1^E \leq Th_1 \cap CD_2^E \leq Th_2 \cap CD_3^E \leq Th_3 \cap CD_4^E \leq Th_4\}. \quad k = 4.$$

where superscript E denotes the inter-class comparisons. The final fused FAR for $k = 1$ is computed by counting CD_i^E less than or equal to Th_i denoted as $C(CD_i^E, Th_i)$

$$FAR = \sum_{i=1}^{n} \frac{C(CD_i^E \leq Th_i)}{Q_i} \tag{2.22}$$

where Q is the total number of inter-class comparisons. Similarly, final FARs are obtained for the different values of $k = 2, 3, 4$ (k of n-possible combinations). Through this process, the final fused FARs and FRRs are obtained separately for each value of k (fused ROC for each k from n-multiple ROCs).

2.5 Experimental Results

This section evaluates the proposed approach using UBIRIS [42], MMU1 [43], CASIA-IrisV2.0 (device1) [44], CASIA-IrisV3 (Interval) [44], and IITD [45] databases. The details of these databases have been provided in Appendix A. It is observed that segmentation on some of the images of all the databases is not accurate due to noncircular boundaries and poor transition from iris to sclera. Inaccurate segmented iris images are also used for the experimentation. Figure 2.7 shows few samples of inaccurately segmented images which were used for the experimentations. The combination of THFB and k-out-of-n:A has been compared with four successful existing iris recognition algorithms [1, 13–15]. In order to assess the recognition accuracy of fused post-classifier, the performance of THFB is compared for four different values of k.

In this chapter, to test the robustness of the proposed approach in the presence of artifacts, no preprocessing technique is used to isolate artifacts during iris recognition process. The scale and shift invariance are achieved by the normalization and

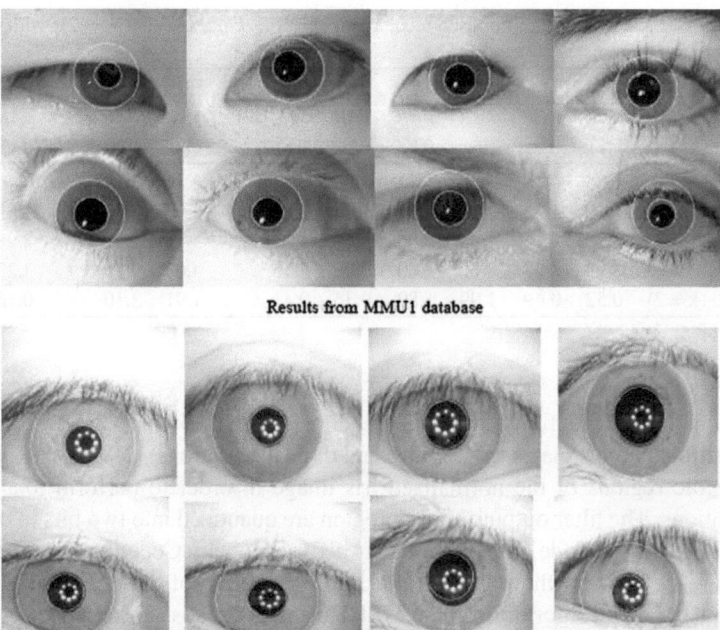

Results from MMU1 database

Result from CASIA-IrisV3-Interval (R) database

Fig. 2.7 Some inaccurate segmented images on MMU1 and CASIA-IrisV3-Interval databases

segmentation processes respectively. The registration process used two iris images per subject. To achieve rotation invariance, five normalized images corresponding to angles -10^0, -5^0, 0^0, 5^0, 10^0 are obtained from each image and used for training. Thus total number of enrolled images for each person is 10. For the testing, remaining three iris images per class are used. In no case images in training and testing sets are overlapping. In order to minimize the effect of intra-class variations and avail the discriminating iris information for the recognition, the upper half iris part from the original normalized iris image is selected. The selected half iris is partitioned into six sub-images and used four regions for the further processing. Each of these four regions is decomposed up-to two levels using these designed filters to create the four feature vectors separately. The artifacts present in iris images lead to reduction in accuracy. In order to improve the performance (reduce the error rate), *k-out-of-n:A* post-classifier is used on these four feature vectors. The performance of the proposed method (combination of THFB and *2-out-of-n:A*) has been compared with four existing well known iris recognition algorithms. These include Daugman [1], Monro et al. [13], Sun and Tan [14], and Dong et al. [15]. These algorithms are implemented and tested on the same set of normalized iris images for comparing the performance of this approach.

The first popular existing algorithm implemented for the comparison with the proposed approaches is the Daugman's algorithm [1]. In this algorithm, Gabor filters

Table 2.2 Comparison of the proposed technique (THFB + k-out-of-n:A ($k = 2$)) with existing iris recognition systems

Algorithms	UBIRIS		MMU1		CAS-IrisV2.0		CAS-IrisV3.0		IITD	
	FAR (%)	FRR (%)	FAR (%)	FRR (%)	FAR (%)	FRR (%)	FAR (%)	FRR (%)	FAR (%)	FRR (%)
Daugman	0.85	0.98	1.35	1.51	0.56	0.67	2.10	2.36	0.46	0.52
Sun and Tan	1.20	1.33	2.72	3.33	1.84	2.13	2.96	3.02	1.35	1.45
Monro et al.	2.29	3.11	4.64	5.56	3.59	4.27	5.16	5.11	2.98	3.01
Dong et al.	2.95	3.28	4.68	5.00	2.86	4.15	4.67	4.89	3.23	3.80
Proposed ($k=2$)	0.52	0.49	1.99	1.89	0.36	0.41	1.91	2.10	0.16	0.15

are parameterized with four degrees-of-freedom: size of the kernels, orientations, and two positional co-ordinates (four scales and three orientations). They are applied on the specific regions of the normalized iris image in order to perform total 1,024 convolutions. The filter outputs of each region are quantized into two bits in order to obtain 2048 bits IrisCode. The dissimilarity between two IrisCodes is estimated using Hamming distance (without using a mask for separation of the region affected by artifacts). In second algorithm, dipole MLDF using 2-D Gaussian kernel (size—5×5 and $\sigma = 1.7$) [14] with $d = 5$ without using bootstrap method was implemented. The extension of this approach as suggested by Dong et al. [15], where the concept of personalized weight map is used for the robustness of encoding algorithm on different iris regions was implemented. It is observed that the robustness of the weight map depends upon the number of training images within a class. If number of training images within a class is more, the performance is significantly improved. This algorithm was trained using three training images per class. The next algorithm implemented is based on DCT coefficients and proposed in [13]. The experimental results for comparison of these algorithms have been presented in Table 2.2.

It is observed from Table 2.2 that the suggested method yields superior performance against these existing iris recognition methods. This is because it works on each selected region of iris independently, so artifacts can only affect the corresponding region and not the entire iris signature. The transformation on iris partitioned sub-regions does not corrupt the good iris region by combining them with artifacts (segmentation error, eyelids /eyelashes occluded regions, etc.). Thus, this approach achieves the robustness to intra-class variations (especially occlusion of pupil on iris due to inaccurate pupil segmentation, occlusion of eyelids/ eyelashes, specular reflection, etc.) in iris recognition system. Figure 2.8a–e shows the comparison in the form of ROC curve of the proposed method with state-of-the-art algorithms on UBIRIS, MMU1, CASIA-IrisV2.0 (device1), CASIA-IrisV3-Inerval, and IITD databases respectively. The influence of k-out-of-n:A post-classifier on the recognition performance of the suggested method was tested for different values of k. The experimental results are shown in Table 2.3.

It is observed that *2-out-of-4:A* post-classifier (acceptance of any two regions out of 4-regions) best suited for all the databases in order to improve the recognition

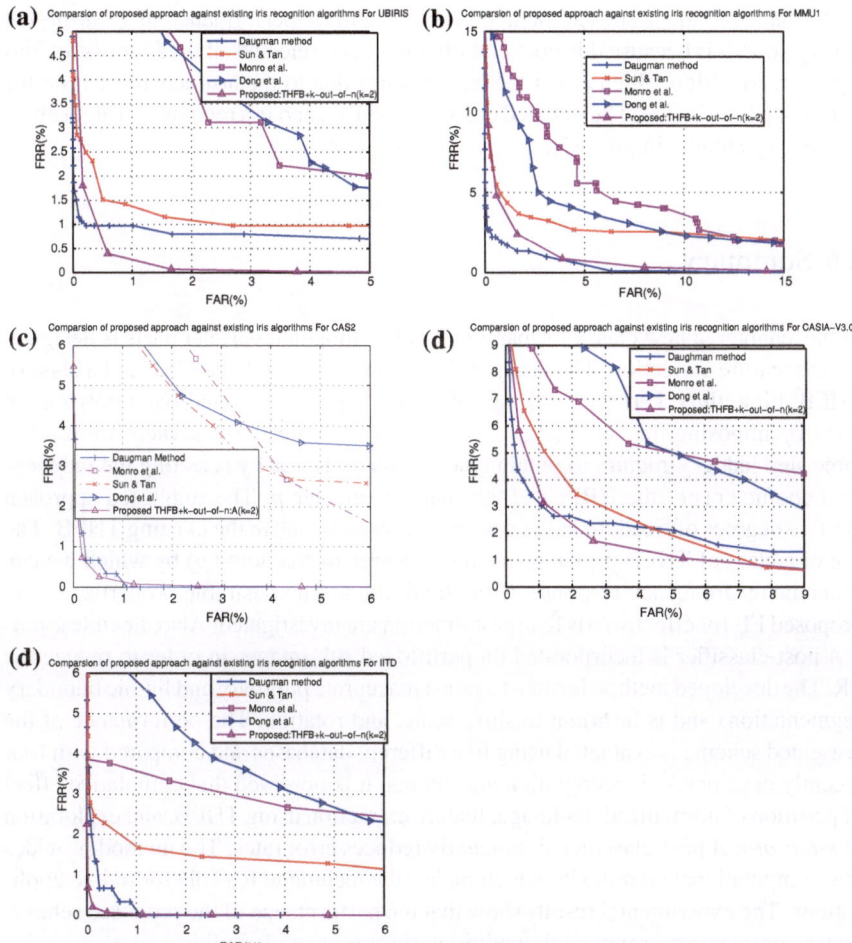

Fig. 2.8 Comparison of proposed algorithm with existing iris recognition algorithms using **a** UBIRIS **b** MMU1 **c** CASIA-IrisV2.0 **d** CASIA-IrisV3.0 **e** IITD iris databases

Table 2.3 Performance of proposed technique (THFB + *k-out-of-n:A*) for different values of k with $n = 4$

THFB +	UBIRIS		MMU1		CAS-IrisV2.0		CAS-IrisV3.0		IITD	
k-out-of-n:A	FAR (%)	FRR (%)	FAR (%)	FRR (%)	FAR (%)	FRR (%)	FAR (%)	FRR (%)	FAR (%)	FRR (%)
k = 1	2.47	2.76	4.45	4.56	1.96	2.03	2.47	2.76	0.92	0.98
k = 2	0.52	0.49	1.99	1.88	0.36	0.41	1.91	2.10	0.16	0.15
k = 3	0.48	0.53	2.38	2.61	0.42	0.46	1.48	1.53	0.30	0.35
k = 4	12.01	13.69	12.07	12.71	3.36	3.30	13.46	13.69	7.51	7.77

performance of the proposed method. The recognition performance of *4-out-of-4:A* is very poor. It is because this post-classifier has been selected all the four iris regions for the recognition which affect the performance due to the artifacts present on the normalized iris image. The detailed design of the FBs and iris recognition algorithms is given by authors in [46, 47].

2.6 Summary

In this chapter, a new class of separable 2-D biorthogonal wavelet basis is designed for iris feature extraction. The design is based on the generalized HBP and a class of THFB. First, three kernels are designed from the generalized halfband polynomial $P(z)$ by imposing the zeros at $z = -1$. These three designed kernels are used in three step ladder structure to obtain a new class of THFB by selecting the independent coefficients of the HBPs and the shape parameter p. The suggested approach has two degrees of freedom in the design compared to one in the existing THFB. The use of a class of THFB provides one more degree of freedom (p) by which we can shape better frequency response of the final filters. The desirable properties of the proposed FB for effective iris feature extraction are investigated. Also, flexible *k-out-n:A* post-classifier is incorporated on partitioned sub-images in order to reduce the FR. The developed method is robust against inaccurate pupillary and limbic boundary segmentations and is invariant to shift, scale, and rotation. The performance of the presented scheme is evaluated using five different databases and compared with four recently developed iris recognition algorithms. It is observed that cumulative effect of partition of normalized iris image, feature extraction using THFB, and exploration of *k-out-of-n:A* post-classifier significantly reduces error rates. The method provides low computational complexity which makes the technique feasible for online applications. The experimental results show that the performance of the proposed scheme under non-ideal environmental conditions (in presence of eyelids/eyelashes occlusion, inaccurate segmentation of inner and outer iris boundaries, specular reflection, etc.) is superior to recently developed iris recognition algorithms.

References

1. Daugman JG (1993) High confidence visual recognition of persons by a test of statistical independence. IEEE Trans Pattern Anal Mach Intell 25(11):1148–1161
2. Proença H, Alexandre LA (2007) Toward non-cooperative iris recognition: a classification approach using multiple signatures. IEEE Trans Pattern Anal Mach Intell 29(4):607–612
3. Masek L (2003) Recognition of human iris patterns for biometric identification. Master's thesis, Department of Computer Science and Software Engineering, The University of Western Australia
4. Vatsa M, Singh R, Noore A (2008) Improving iris recognition performance using segmentation, quality enhancement, match score fusion, and indexing. IEEE Trans Syst Man Cyber B

   38(4):1021–1035
5. Boles WW, Boashash B (1998) A human identification technique using images of the iris and wavelet transform. IEEE Trans Signal Process 46(4):1185–1188
6. Wildes RP (1997) Iris recognition: an emerging biometric technology. Proc IEEE 85(9): 1348–1363
7. Lim S, Lee K, Byeon O, Kim T (2001) Efficient iris recognition through improvement of feture vector and classifier. ETRI 23((2)):61–70
8. Ma L, Tan T, Wang Y, Zhang D (2004) Personal identification based on iris texture analysis. IEEE Trans Pattern Anal Mach Intell 25(12):2519–2533
9. Ma L, Wang Y, Zhang D (2004) Efficient iris recognition by characterizing key local variations. IEEE Trans Image Process 13(6):739–750
10. Nabti M, Ghouti L, Bouridane A (2008) An effective and fast iris recognition system based on a combined multiscale feature extraction technique. Pattern Recogn 41:868–879
11. Velisavljević V (2009) Low-complexity iris coding and recognition based on directionlets. IEEE Trans Inf Forensic Secur 4(3):410–417
12. Abhyankar A, Schuckers S (2007) A novel biorthogonal wavelet network system for off-angle iris recognition. Pattern Recogn 43:586–594
13. Monro DM, Rakshit S, Zhang D (2007) DCT based iris recognition. IEEE Trans Pattern Anal Mach Intell 29(4):586–595
14. Sun Z, Tan T (2009) Ordinal measures for iris recognition. IEEE Trans Pattern Anal Mach Intell 31(12):2211–2226
15. Dong W, Tan T, Sun Z (2010) Iris matching based on personalized weight map. IEEE Trans Pattern Anal Mach Intell 99(1):1–14
16. Kumar A, Passi A (2010) Comparison and combination of iris matchers for reliable personal authentication. Pattern Recogn 43(3):1016–1026
17. Huang J, You X, Yuan Y, Yang F, Lin L (2010) Rotation invariant iris feature extraction using Gaussian Markov random fileds with non-seprable wavelet. Neurocomputing 73:883–894
18. Park C, Lee J, Oh S, Song Y (2003) Iris feature extraction and matching based on maltiscale and directional image representation. In: Proceedings of 4th international conference on scale space methods in computer vision, Isle of Skye, UK, 2003, Springer-Verlag, pp 576–583
19. Patil B, Patwardhan P, Gadre V (2008) On the design of fir wavelet filter banks using factorization of a halfband polynomial. IEEE Signal Process Lett 15:485–488
20. Daubechies I (1992) Ten lectures on wavelets. SIAM, Philadelphia
21. Daubechies I, Feauveau J (1992) Biorthogonal bases of compactly supported wavelets. Commun Pure Appl Math 45:485–560
22. Ansari R, Kaiser CGJ (1991) Wavelet construction using Lagrange halfband filters. IEEE Trans Circuits Syst Express Brief 38(9):1116–1118
23. Sweldens W (1996) The lifting scheme: a custom-design construction of biorthogonal wavelets. Appl Comput Harmonic Anal 3(2):186–200
24. Phoong S, Kim C, Vaidyanathan P, Ansari R (1995) A new class of two-channel biorthogonal filter banks and wavelet bases. IEEE Trans Signal Process 43(3):649–665
25. Ansari R, Kim C, Dedovic M (1999) Structure and design of two-channel filter banks derived from a triplet of halfband filters. IEEE Trans Circuits Syst II Analog Digital Signal Process 46(12):1487–1496
26. Tay DBH, Palaniswami M (2004) A novel approach to the design of the class of triplet halfband filterbanks. IEEE Trans Circuits Syst Syst II Express Brief 51(7):378–383
27. Tay DBH (2008) ETHFB: A new class of even-length biorthogonal wavelet filters for Hilbert pair design. IEEE Trans Circuits Syst Syst I Regul Pap 55(6):1580–1588
28. Chan S, Yeung K (2004) On the design and multiplierless realizaton of perfect reconstruction triplet-based fir filterbanks and wavelet bases. IEEE Trans Circuits Syst Syst I 51(8): 1476–1491
29. Kha H, Tuan H, Nguyen T (2011) Optimal design of fir triplet halfband filter bank and application in image coding. IEEE Trans Image Process 22(2):586–591

30. Eslami R, Radha H (2010) Design of regular wavelets using a three-step lifting scheme. IEEE Trans Signal Process 58(4):2088–2101
31. Kovacevic J, Sweldens W (2000) Wavelet families of increasing order in arbitrary dimensions. IEEE Trans Image Process 9(3):480–496
32. Strang G, Nguyen T (1996) Wavelets and filter banks. Wellesley-Cambridge, NY
33. Vetterli M, Kovacevic J (1995) Wavelets and subband coding. Prentice-Hall, Englewood Cliffs
34. Mojsilovic A, Popovic M, Rackov D (2008) On the selecton of optimal wavelet basis for texture characterization. IEEE Trans Image Process 9(12):2043–2050
35. Vaidyanathan PP (1993) Multirate systems and filter banks. Prentice-Hall, Englewood Cliffs
36. Tay DBH (2001) Balanced spatial and frequency localized 2-D nonseparable wavelet filters. In: Proceedings of IEEE international symposium on circutis and systems, May 2001, vol 2, pp 489–492
37. Monro D, Sherlock B (1997) Space frequency balance in biorthogonal wavelets. In: Proceedings of IEEE international conference on image processing, October 1997, vol 1, pp 642–627
38. Chen Y, Dass S, Jain A (2006) Localized iris image quality using 2-D wavelets. In: Proceedings of international conference on biometrics, 2006, pp 373–381
39. Ross A, Jain A (2003) Information fusion in biometrics. Pattern Recogn Lett 24(13): 2115–2125
40. Daugman JG (2000) Biometric decision landscapes. University of Cambridge, Technical Report
41. Ebeling E (1997) An introduction to reliability and maintability engineering. McGraw-Hill, International edition, New York
42. Proenca H, Alexandre L. UBIRIS: a noisy iris image database. www.iris.di.ubi.pt
43. Multimedia University (2004) MMU iris image database. http://pesona.mmu.edu.my/ccteo
44. CASIA iris image database. http://www.sinobiometrics.com/casiairis.htm
45. IITD iris image database. http://web.iitd.ac.in
46. Rahulkar AD, Holambe RS (2012) Half-iris feature extraction and recognition using a new class of biorhtogonal triplet half-band filter band and flexible k-out-of-n:A. Postclassifier. IEEE Trans Inf Forensic Secur 7(1):230–240
47. Rahulkar AD, Patil BD, Holambe RS (2012) A new approach to the design of biorthogonal triplet halfband filter banks using generalized halfband polynomials. Signal Image Video Process (Springer) 1–12

Chapter 3
Combined Directional Wavelet Filter-Banks Based Features

Abstract This chapter presents the construction of combined directional wavelet filter banks (CDWFB) by the combination of directional wavelet filter bank (DWFB) and rotated directional wavelet filter banks. The iris image feature extraction algorithm based on CDWFB and post-classifier have been discussed in this chapter.

Keywords Checkerboard shaped filter bank · Directional wavelet filter banks · Fan shaped filter banks · Rotated directional wavelet filter banks · Wavelets

3.1 Introduction

The iris recognition algorithm presented in Chap. 2 suffered from limited directionality (i.e. three directions-horizontal, vertical, and mixed diagonal) and also contained mixed diagonal information. This algorithm extracts only three directional iris information (limited information) which may fail to perform well in presence of artifacts. In this chapter, we present an iris recognition algorithm based on combined directional wavelet filter bank (CDWFB) that provides twelve directional iris information.

3.2 Review of the Related Directional Filter Bank

Psychophysics and physiological experiments have shown that multiscale transforms seem to appear in the visual cortex of mammals [1]. The research in HVS shows that directional information plays an important role in visual perception. This result supports the hypothesis that the HVS has been tuned to extract the essential information of a natural scene using a least number of visual active cells [2]. This is an important motivation to derive efficient computational iris texture which should based on multi-resolution, local, directional, and critical sampling expansion.

This chapter presents design of the non-separable FBs in iris recognition system so as to derive compact directional iris features. Patil et al. [3] presented a novel approach to obtain the FBs with flexible frequency response by the factorization of a

HBP. Kim and Udpa [4] introduced a set of rotated wavelet filters (RWFs) using Haar wavelet coefficients to separate the diagonal information at 45° and −45°. However, this approach has limited directionality. Kokare et al. [5] constructed rotated complex wavelet filters (RCWFs) by complex wavelet filter coefficients to overcome the limitations of wavelet filters. The work presented by [5] used the same approach as given in [4]. This set of RCWFs is two times redundant than wavelet filters. Eslami and Radha [6] constructed a new family of non-redundant hybrid geometrical image transform using standard 9/7 BWFB and modified DFBs. This modified DFB shows good results in image coding and denoising at the cost of slow frequency roll-off. The objective of Lu and Do [7] was to provide the wavelet FB with finer directionality. In their work, checkerboard shaped filter bank (CSFB) is constructed based on the modified parameterizations of the polyphase matrices. This form of parameterizations is suggested by Phoong et al. [8] which has some inherent restrictions (see [9]).

The main contributions of this chapter are summarized as follows: First, 2-D separable wavelet FBs are obtained by designing 1-D filter coefficients. Secondly, a pair of fan shaped filter bank (FSFB) is designed by McClellan transformation on the designed 1-D filter coefficients. This pair of FSFB is passed through quincunx sampling matrix to obtain a pair of CSFB. The pair of CSFB is applied on the horizontal, vertical, and mixed diagonal sub-bands of the separable wavelet filters to obtain two mostly horizontal, two mostly vertical, and two separate diagonal directional filters. With this, directional wavelet filter bank (DWFB) is obtained that provides one low-pass sub-band and six finer directional high-pass sub-bands at each scale. This DWFB rotated by an angle of 45° to obtain rotated directional wavelet filter bank (RDWFB) that gives six different directions. Thus, the proposed design yields total 12-directional information by combining DWFB and RDWFB (CDWFB) to extract the iris features. Furthermore, k-out-of-n:A post-classifier (explained in Sect. 2.4.2) is used to handle possible artifacts especially pupil segmentation error, occlusion of eyelids/eyelashes, reflection on iris, non-linear deformation, etc.

3.3 Construction of the Directional Filter Bank

3.3.1 Design of 1-D Biorthogonal Wavelet FB Using Factorization of an HBP

Design of biorthogonal wavelet filter bank (BWFB) by the factorization of a generalized HBP [3] has been discussed in Sect. 2.2.2. In this chapter, 18th order HBP is considered to design 1-D BWFB. The FB design from 18th order generalized HBP is as follows:

$$P(z) = a_0 + a_2 z^{-2} + a_4 z^{-4} + a_6 z^{-6} + a_8 z^{-8} + z^{-9}$$
$$+ a_8 z^{-10} + a_6 z^{-12} + a_4 z^{-14} + a_2 z^{-16} + a_0 z^{-18}.$$

Synthetic-division is used to factorize this P(z) by imposing eight zeros at $z = -1$. Imposing the flatness gives the following set of constraints:

$$a_8 = \frac{1}{2} - a_0 - a_2 - a_4 - a_6; \quad a_6 = -10a_0 - 6a_2 - 3a_4 - \frac{1}{16}$$
$$a_4 = -15a_0 - 5a_2 + \frac{3}{256}; \quad a_2 = -7a_0 - \frac{5}{2048}.$$

These constraints are used in $P(z)$ to obtain

$$P(z) = (1 + z^{-1})^8 R(z).$$

where $R(z)$ is the remainder term (is a function of parameter a_0) obtained as follows:

$$R(z) = a_0 - 8a_0 z^{-1} + \left(\frac{-160}{65536} + 29a_0\right) z^{-2} + \left(\frac{1280}{65536} - 64a_0\right) z^{-3}$$
$$+ \left(\frac{-4192}{655360} + 98a_0\right) z^{-4} + \left(\frac{6656}{65536} - 122a_0\right) z^{-5}$$
$$+ \left(\frac{-4192}{65536} + 98a_0\right) z^{-6} + \left(\frac{1280}{65535} - 64a_0\right) z^{-7}$$
$$+ \left(\frac{-160}{65536} + 29a_0\right) z^{-8} - 8a_0 z^{-9} + a_0 z^{-10}.$$

Now factorized this $R(z)$ into $R_1(z)$ and $R_2(z)$ for the optimized value of $a_0 = 0.000085198$ (MATLAB optimization routine *fminunc* is used to choose value of a_0) such that energy of ripples are minimized. The factorization of $R(z)$ into $R_1(z)$ and $R_2(z)$ is given as:

$$R_1(z) = 0.1250 - 0.4409z^{-1} - 3.1067z^{-2} + 11.47z^{-3}$$
$$- 3.1067z^{-4} - 0.4409z^{-5} + 0.1250z^{-6},$$
$$R_2(z) = 0.25 - 1.1181z^{-1} + 2.355z^{-2} - 1.1181z^{-3} + 0.25z^{-4}.$$

The flexibility in frequency response of the filters (regularity) $H_0(z)$ and $G_0(z)$ can be obtained by the re-assignment of eight zeros to above factors. In this work, four zeros are assigned to each factor $R_1(z)$ and $R_2(z)$. Thus, the final designed (11/9) filters are:

$$H_0(z) = (1 + z^{-1})^4 R_1(z),$$
$$G_0(z) = (1 + z^{-1})^4 R_2(z).$$

The length of $H_0(z)$ and $G_0(z)$ is 11 and 9, respectively, so called 11/9 filters. The analysis and synthesis HPFs are obtained with the help of quadrature mirroring the LPF coefficients. The frequency responses of the analysis LP and HP filters are shown in Fig. 3.1.

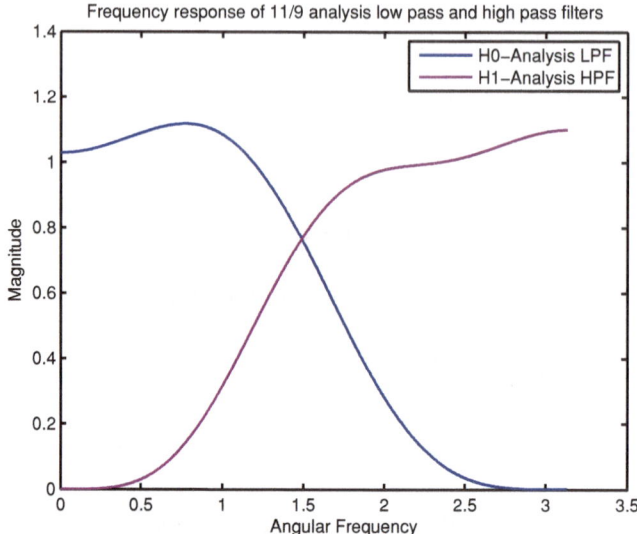

Fig. 3.1 Frequency responses of the designed 11/9 analysis LP and HP filters

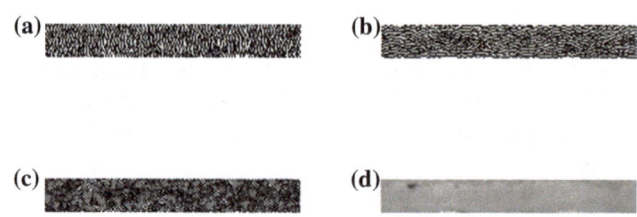

Fig. 3.2 One-level decomposition of CASIA-IrisV3.0 iris image using 11/9 FB **a** LH sub-band **b** HL sub-band **c** HH sub-band **d** approximate (LL) sub-band

3.3.2 Construction of 2-D Separable Filter Bank

The three detail and one approximation sub-bands obtained from the 11/9 LPF and HPF coefficients are as follows:

$$
\begin{aligned}
L(z) &= H_0^{1d}(z_1) \times H_0^{1d}(z_2), \\
H(z) &= H_0^{1d}(z_1) \times H_1^{1d}(z_2), \\
V(z) &= H_1^{1d}(z_1) \times H_0^{1d}(z_2), \\
D(z) &= H_1^{1d}(z_1) \times H_1^{1d}(z_2).
\end{aligned}
\tag{3.1}
$$

where $H_0^{1d}(z_1)$ and $H_1^{1d}(z_1)$ are the designed 11/9 1-D LPF and HPF respectively. The one-level decomposition results in *Vertical* (V), *Horizontal* (H), and *Diagonal* (D) sub-bands and one *approximation sub-band* (L), which corresponds to LH, HL, HH, and LL sub-bands respectively as shown in Fig. 3.2a–d for one of the iris images.

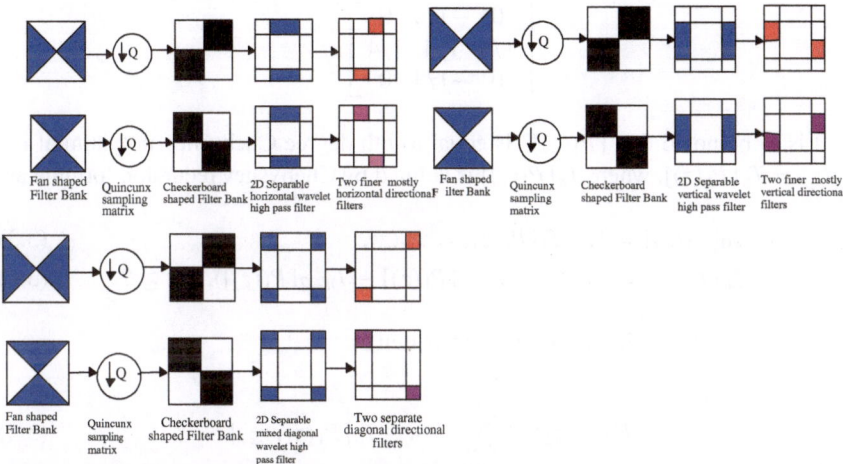

Fan shaped Filter Bank   Quincunx sampling matrix   Checkerboard shaped Filter Bank   2D Separable horizontal wavelet high pass filter   Two finer mostly horizontal directional filters   Fan shaped Filter Bank   Quincunx sampling matrix   Checkerboard shaped Filter Bank   2D Separable vertical wavelet high pass filter   Two finer mostly vertical directional filters

Fan shaped Filter Bank   Quincunx sampling matrix   Checkerboard shaped Filter Bank   2D Separable mixed diagonal wavelet high pass filter   Two separate diagonal directional filters

Fig. 3.3 Proposed cascade directional wavelet FB structure

It is observed that HH sub-band of Fig. 3.2c contained mixed two diagonal orientations (+45° and −45°). Consequently, separable wavelet FBs fail to separate these orientations. Also, separable wavelet FBs provide only three-directional information. A number of approaches are presented in the literature to provide more directional information. Many researchers have shown that multidimensional (MD) non-separable filtering via McClellan transformation is a useful technique for designing linear-phase MD filters. The design of MD non-separable FBs using mapping of 1-D frequencies into isopotentials (contours in the 2-D frequency plane) is presented in [10, 11]. In this work, each HP sub-band of designed separable FB is partitioned into two parts using a pair of CSFB to obtain the finer directions. The proposed cascade structure is shown in Fig. 3.3.

3.3.3 Construction of Fan Shaped Filter Bank

The McClellan transformation is used on the designed 1-D filter coefficients to obtain non-separable Fan Shaped Filter Bank (FSFB). The requirements of the McClellan transformation are:

- Zero-phase 1-D filter coefficients.
- Zero-phase transformation kernel.

McClellan transformation is expressed as follows [12]:

$$H(z) = \sum_{n=0}^{N} h(n)[P_1(z)]^n. \tag{3.2}$$

where $P_1(z) = ((z^1 + z^{-1})/2)$, and N is the order of zero-phase 1-D filter $h(n)$. The following transformation kernel is used to obtain FSFB.

$$\begin{bmatrix} 0 & -1/4 & 0 \\ 1/4 & 1/4 & 1/4 \\ 0 & -1/4 & 0 \end{bmatrix}.$$

It is to be noted that $[P_1(z)]^n$ is equal to nth degree Chebyshev polynomial i.e. $P_n(z) = T_n[P_1(z)]$, where $T_n[P_1(z)]$ is defined by Chebyshev recursion formula as:

$$T_0[P_1(z)] = 1, \quad T_1[P_1(z)] = P_1(z), \tag{3.3}$$

$$T_n[P_1(z)] = 2P_1(z) \cdot (T_{n-1}[P_1(z)] - T_{n-2}[P_1(z)]), \quad n \geq 2. \tag{3.4}$$

Using this assertion, Eq. (3.2) can be rewritten as:

$$H(z_1, z_2) = \sum_{n=0}^{N} 2h(n) \cdot T_n[P_1(z_1, z_2)]. \tag{3.5}$$

Thus, a pair of FSFB is obtained as:

$$F_{0i}(z_1, z_2) = \sum_{n=0}^{N} 2h_i(n) \cdot T_n[P_1(z_1, z_2)], \quad i \epsilon[0, 1]. \tag{3.6}$$

This fan shaped filters $F_0(z_1, z_2)$ and $F_1(z_1, z_2)$ are sampled by the quincunx sampling matrix to construct a pair of CSFB using Eq. (3.7) as:

$$\begin{aligned} C_0(z_1, z_2) &= F_0(z_1 z_2, z_1 z_2^{-1}), \\ C_1(z_1, z_2) &= F_1(z_1 z_2, z_1 z_2^{-1}). \end{aligned} \tag{3.7}$$

The frequency responses of a pair of FSFB $F_0(z_1, z_2)$ and $F_1(z_1, z_2)$ and CSFB $C_0(z_1, z_2)$ and $C_1(z_1, z_2)$ are shown in Figs. 3.4 and 3.5 respectively.

3.3.4 Construction of Directional Wavelet Filter Bank

The three detail and one approximate sub-bands are obtained by 2-D 11/9 FB at the first level of Directional Wavelet Filter Bank (DWFB). During the second level, CSFB is applied on each detail sub-band of 2-D 11/9 FB to obtain six different finer directional information using Eq. (3.8) as:

$$\begin{aligned} H_{00}(z_1, z_2) &= H(z_1, z_2)C_0(z_1, z_2), \\ H_{01}(z_1, z_2) &= H(z_1, z_2)C_1(z_1, z_2), \\ H_{02}(z_1, z_2) &= V(z_1, z_2)C_0(z_1, z_2), \\ H_{03}(z_1, z_2) &= V(z_1, z_2)C_1(z_1, z_2), \\ H_{04}(z_1, z_2) &= D(z_1, z_2)C_0(z_1, z_2), \\ H_{05}(z_1, z_2) &= D(z_1, z_2)C_1(z_1, z_2). \end{aligned} \tag{3.8}$$

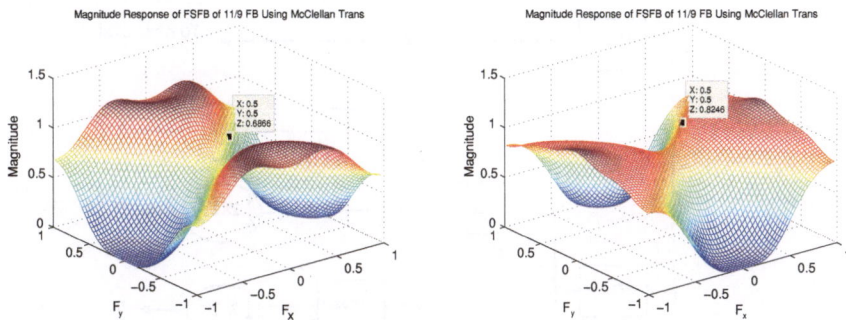

Fig. 3.4 Frequency responses of the pair of FSFB

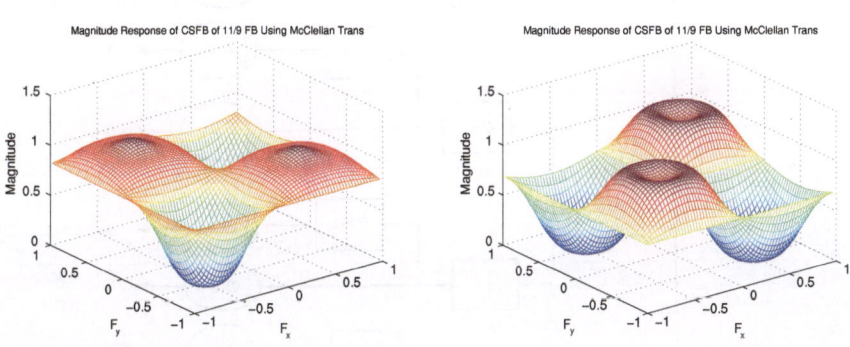

Fig. 3.5 Frequency responses of the pair of CSFB

Thus, six directional sub-bands are obtained which are approximately oriented in $\theta_1 = (15°, 45°, 75°, 105°, 135°, 165°)$ directions. It is observed that DT-CWT [13] also provides the same directional information. However, DT-CWT is four times redundant whereas the proposed approach is non-redundant due to the usage of critical sampling.

3.3.5 Construction of Rotated Directional Wavelet Filter Bank

Rotated Directional Wavelet Filter Bank (RDWFB) is obtained by rotating DWFB with an angle of 45°. The decomposition is performed along the new directions which are 45° apart from the decomposition directions of standard wavelet filters. The 2-D filters i.e. $h_{000}, h_{001}, h_{002}, h_{003}, h_{004}, h_{005}$ are 45° rotated versions of the DWFB as shown in Fig. 3.6. The synthetically generated image is decomposed using DWFB and RDWFB up to one-level in order to visualize the corresponding directional information as shown in Fig. 3.7.

It is observed that DWFB gives the information approximately oriented in $\theta_1 = (15°, 45°, 75°, 105°, 135°, 165°)$ and RDWFB provides the information approximately oriented in $\theta_2 = (0°, 30°, 60°, 90°, 120°, 150°)$. The directional information

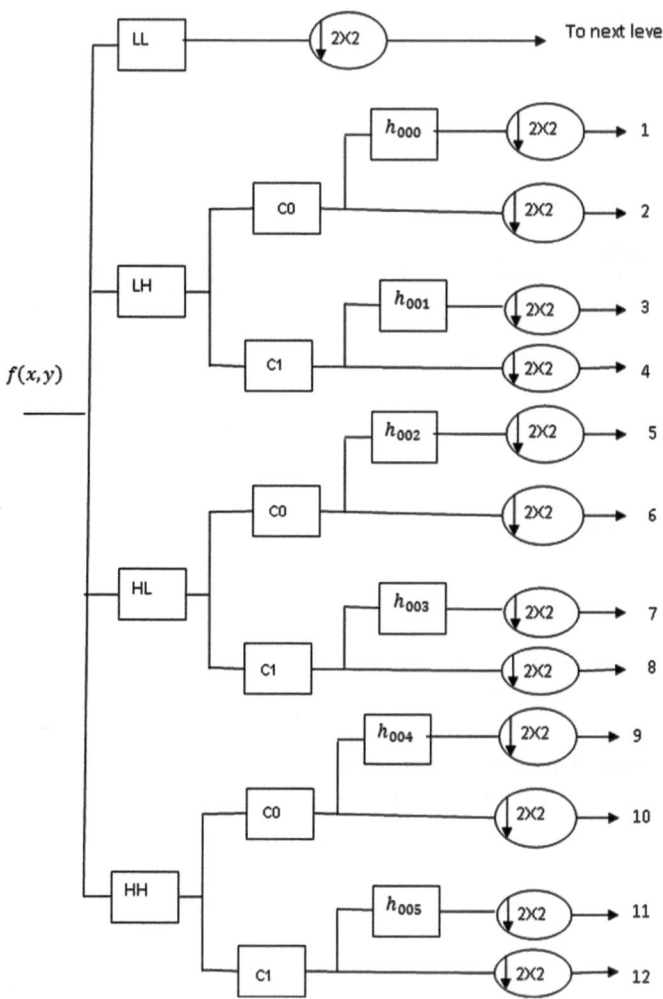

Fig. 3.6 Filter bank construction of the proposed CDWFB

of DWFB and RDWFB are combined together (CDWFB) to obtain 12-directional information as $\theta = (0°, 15°, 30°, 45°, 60°, 75°, 90°, 105°, 120°, 135°, 150°, 165°)$. Since, each individual filter of 11/9 FB is critically sampled; the overall FB is also critically sampled. The same set of directions is obtained by RCWFs [5]. However, RCWFs are obtained using known complex wavelet filter coefficients, which are two times redundant than the proposed DWFB.

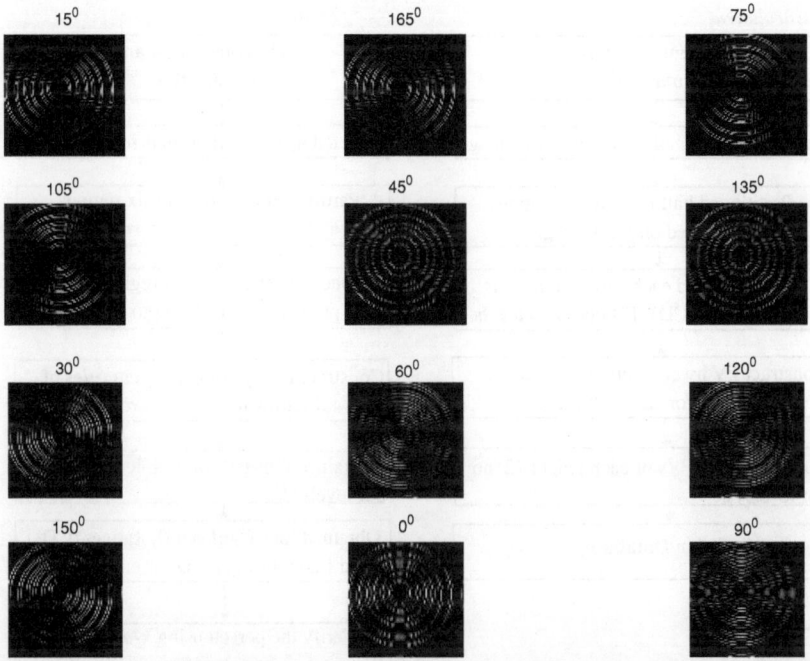

Fig. 3.7 One-level decomposition of image using proposed structure to show 12 directional information

3.4 Feature Extraction Using DWFB and RDWFB

The block diagram of the proposed iris recognition algorithm is shown in Fig. 3.8. In this section, directional iris texture features based on CDWFB are computed. The inner half iris region is divided into six sub-images and selected only four regions for further processing. The CDWFB is applied separately on each of the four selected sub-blocks (sub-images). The feature vector for each sub-image is derived by computing the energies of the CDWFB sub-bands. The four distance scores are obtained and fused using post-classifier in order to improve the recognition performance.

In this chapter, iris is segmented using Daugman's integro-differential operator and normalized by Daugman's rubber sheet model. The preprocessing steps are shown in Fig. 2.6a–c.

Since region closer to the pupil provides more useful iris information, the upper half iris from the entire normalized iris image (inner iris region) is selected. This partial iris is divided into six sub-regions and only four sub-regions are selected (Region: 1, 3, 4, and 6 as shown in Fig. 2.6d) in order to minimize the effect of artifacts during iris recognition process. The occlusion of eyelids/eyelashes are minimum in these selected regions as compared to other two regions (Region: 2 and 5). The

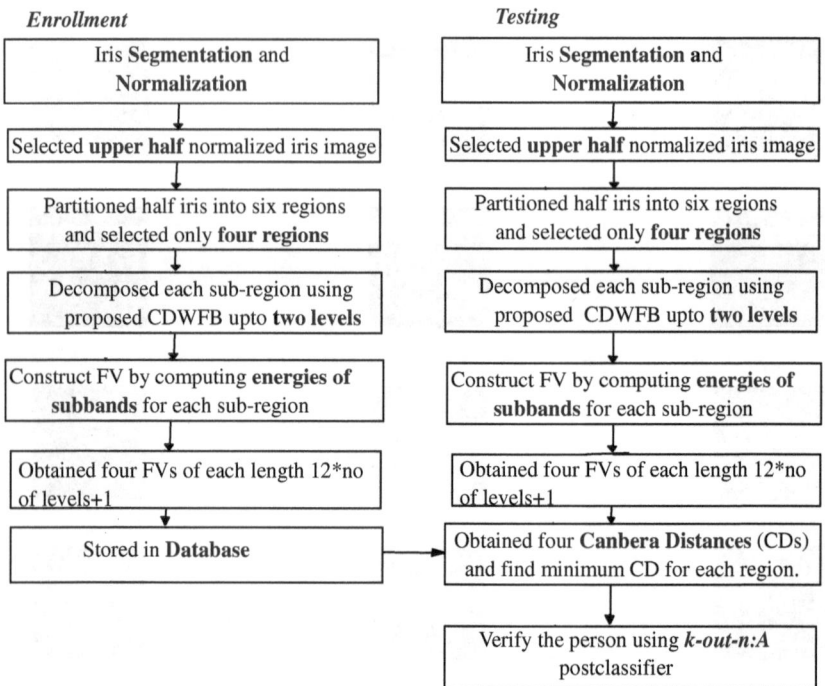

Fig. 3.8 Block diagram of proposed iris recognition algorithm

CDWFB is applied on each of these four regions to extract 12-directional iris texture. Each sub-band in CDWFB represents iris characteristics in a particular scale and direction. As energy is an important characteristic in identifying texture (which is normally being used in the literature to represent textures), the normalized directional energy is computed by L_1 norm from each sub-band of the CDWFB of scale s and direction d as below:

$$E_i = \frac{1}{N_1 \times N_2} \sum_{m=1}^{N_1} \sum_{n=1}^{N_2} |W_{s,d}(m,n)|. \qquad (3.9)$$

where $W_{s,d}(m,n)$ is the set of sub-band coefficients at scale s and direction d, $N_1 \times N_2$ is the total number of coefficients in that sub-band. These features at different scales and orientations are concatenated to derive the feature vector as

$$E = [E_{1,1}, E_{1,2}, E_{1,3}, \ldots, E_{S,D_k}, E_a]. \qquad (3.10)$$

where S and D_k are the total number of scales and total number of orientations at kth scale, E_a is the energy of an approximate sub-band. The total number of sub-bands for CDWFB is $(12 \times S) + 1$. The derived feature vectors of each region are stored

Table 3.1 Comparison of proposed technique (CDWFB + k-out-of-n:A ($k = 2$)) with existing iris recognition algorithms

Algorithms	UBIRIS		MMU1		CAS-IrisV2.0		CAS-IrisV3.0		IITD	
	FAR (%)	FRR (%)	FAR (%)	FRR (%)	FAR (%)	FRR (%)	FAR (%)	FRR (%)	FAR (%)	FRR (%)
Daugman	0.85	0.98	1.35	1.51	0.56	0.67	2.10	2.36	0.46	0.52
Sun and Tan	1.20	1.33	2.72	3.33	1.84	2.13	2.96	3.02	1.35	1.45
Monro et al.	2.29	3.11	4.64	5.56	3.59	4.27	5.16	5.11	2.98	3.01
Dong et al.	2.95	3.28	4.68	5.00	2.86	4.15	4.67	4.89	3.23	3.80
Proposed ($k=2$)	0.20	0.23	0.62	0.70	0.08	0.09	0.76	0.88	0.025	0.029

in the database as reference (enrollment process). The test iris pattern is classified on the basis of minimum Canbera Distance (CD) between test iris feature vector and that of feature vectors stored in the database. The computation of Canbera distance is given in Eq. (2.18).

In order to improve the performance of the proposed scheme, k-out-of-n:A post-classifier is used at the decision level (mentioned in Sect. 2.4.2). Through this process, the final fused FARs and FRRs have been obtained.

3.5 Experimental Results

This section evaluates the suggested approach using UBIRIS [14], MMU1 [15], CASIA-IrisV2.0 (device1) [16], CASIA-IrisV3-Interval [16], and IITD [17] databases. The details of these databases have been given in Appendix A. The suggested approach (combination of CDWFB and k-out-of-n:A) has been compared with four successful existing iris recognition algorithms [18–21]. In order to assess the recognition accuracy of the post-classifier, the performance of CDWFB is compared for four different values of k.

The preprocessing is carried out in the same way as given in Fig. 2.6 for training and testing phases. The selected partial iris is partitioned into six sub-images and used four regions for further processing. Each of these four regions is decomposed up-to two levels using CDWFB to create the four feature vectors and obtained four Canbera Distances separately. In order to achieve robustness against the artifacts (segmentation error, eyelids/eyelashes obstruction, reflections, contrast changes, etc.), k-out-of-n:A post-classifier is used on these four CDs. During training, four independent ROCs are obtained from corresponding inter-class and intra-class comparisons. These ROCs are fused by k-out-of-n:A post-classifier to get a single ROC. From this ROC, optimal operating point is decided by the possible combinations of FAR and FRR. With this optimal operating point, thresholds of corresponding regions are obtained. During the testing, final decision for the authentication is taken based on the trained thresholds. The performance of this method has been compared with

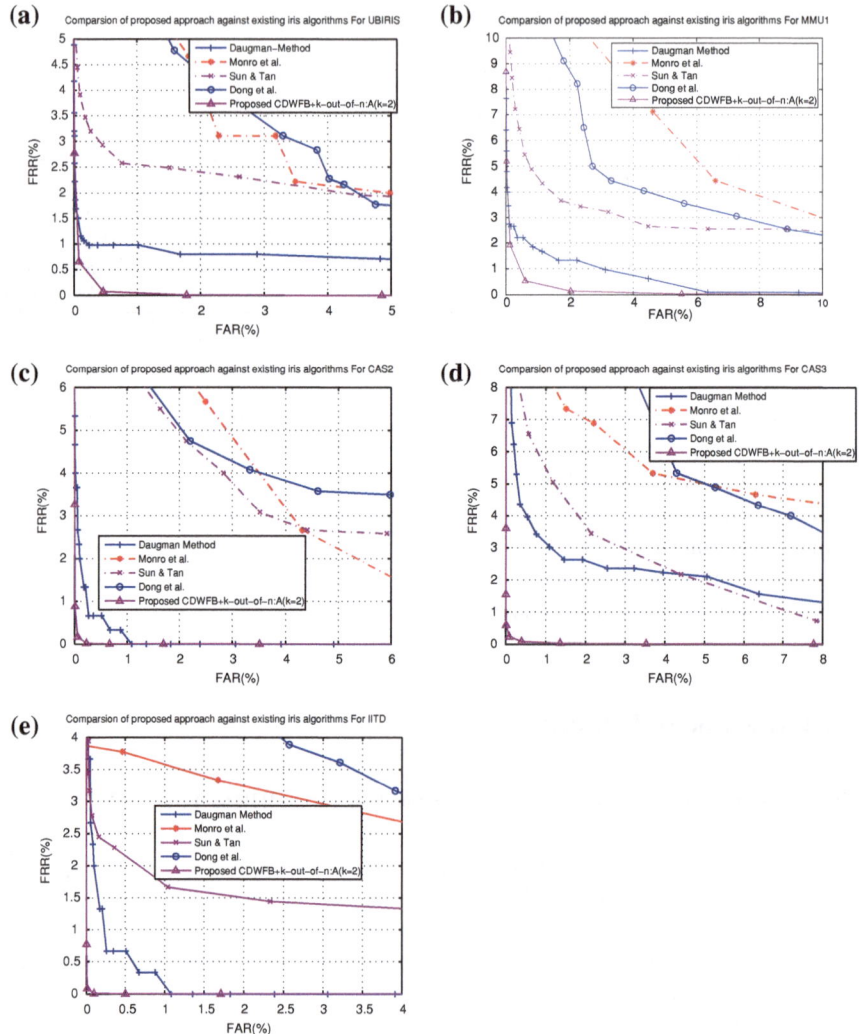

Fig. 3.9 Comparison of proposed algorithm with existing iris recognition algorithms using **a** UBIRIS **b** MMU1 **c** CASIA-IrisV2.0 **d** CASIA-IrisV3.0 **e** IITD iris databases

four recent existing well known iris recognition algorithms. These include Daugman [18], Monro et al. [19], Sun and Tan [20], and Dong et al. [21]. These algorithms are implemented and tested on the same set of normalized iris images for comparing the performance of the proposed approach. The implementation of these existing iris recognition algorithms is given in Sect. 2.5. The experimental results have been presented in Table 3.1. It is observed from Table 3.1 that the proposed method yields superior performance against existing iris recognition methods. This is because the proposed scheme provides more directional iris information and works on each

Table 3.2 Comparison of proposed technique (CDWFB + *k-out-of-n:A*) for different values of *k*

CDWFB + k-out-of-n:A	UBIRIS		MMU1		CAS-IrisV2.0		CAS-IrisV3.0		IITD	
	FAR (%)	FRR (%)	FAR (%)	FRR (%)	FAR (%)	FRR (%)	FAR (%)	FRR (%)	FAR (%)	FRR (%)
k = 1	0.72	0.85	0.93	1.09	0.43	0.51	1.21	1.46	0.22	0.21
k = 2	0.20	0.23	0.62	0.70	0.08	0.09	0.76	0.88	0.025	0.029
k = 3	0.20	0.22	0.93	0.92	0.07	0.089	1.70	2.02	0.06	0.09
k = 4	3.06	3.17	5.97	6.31	1.34	1.35	11.85	12.22	2.66	2.73

region of iris independently, so artifacts can only affect the corresponding region and not the entire iris signature. The transformation on iris partitioned sub-regions does not corrupt the good iris region by combining them with artifacts (segmentation error, eyelids/eyelashes occluded regions, etc.). Thus, this introduced approach achieves the robustness to intra-class iris variations (especially occlusion of pupil on iris due to inaccurate pupil segmentation, occlusion of eyelids/ eyelashes, specular reflection etc.). Figure 3.9a–e shows the comparison of the proposed scheme with existing iris recognition algorithms in the form of ROC curves. The detail can be found in [22].

The influence of *k-out-of-n:A* post-classifier on the recognition performance of the proposed method was tested for different values of k. The results for $k = 1, 2, 3, 4$ and $n = 4$ are shown in Table 3.2.

3.6 Summary

In this chapter, a new approach for the design of 2-D non-separable, non-redundant and multi-resolution CDWFB based on the polynomial factorization and McClellan transformation has been developed. The suggested FB provides 12-directional iris information which overcomes the limitations (limited directionality) of the proposed class of THFB. DT-CWT and RCWFs also provide the same information but they are redundant than this design. A novel iris feature extraction scheme is presented using CDWFB on partitioned upper half normalized iris. A simple *k-out-n:A* classifier is applied at the decision level in order to reduce the FRR. The method is translation, scale and rotation invariant. The performance of this approach is evaluated using five different databases and compared with four state-of-the-art iris recognition algorithms. The experimental results show improvement in performance of the suggested scheme under non-ideal environmental conditions.

References

1. Mallat S (1996) Wavelet for a vision. Proc IEEE 84:604–614
2. Do M, Vetterli M (2005) The contourlet transform: an efficient directional multiresolution image representation. IEEE Trans Image Process 14(12):2091–2106
3. Patil B, Patwardhan P, Gadre V (2008) On the design of fir wavelet filter banks using factorization of a halfband polynomial. IEEE Signal Process Lett 15:485–488
4. Kim N, Udpa S (2000) Texture classification using rotated wavelet filters. IEEE Trans Syst Man Cybern Part A Syst Hum 30(6):847–852
5. Kokare M, Biswas P, Chatterji B (2005) Texture image retrieval using new rotated complex wavelet fitlers. IEEE Trans Syst Man Cybern Part B Cybern 35(6):1168–1178
6. Eslami R, Radha H (2007) A new family of nonredundant transforms using hybrid wavelets and directional filter banks. IEEE Trans Image Process 16(4):1152–1167
7. Lu Y, Do M (2005) The finer directional wavelet transform. In: Proceedings of the IEEE ICASSP, Philadelphia, 2005
8. Phoong S, Kim C, Vaidyanathan P, Ansari R (1995) A new class of two-channel biorthogonal filter banks and wavelet bases. IEEE Trans Signal Process 43(3):649–665
9. Ansari R, Kim C, Dedovic M (1999) Structure and design of two-channel filter banks derived from a triplet of halfband filters. IEEE Trans Circ Syst II Analog Digital Signal Process 46(12):1487–1496
10. Kovacevic J, Vetterli M (1992) Nonseparable multidimensional perfect reconstruction filter banks and wavelet bases of r^n. IEEE Trans Inf Theor 38(2):533–555
11. Shah I, Kalker T (1993) Theory and design of multidimensional qmf sub-band filters from 1-d filters and polynomials using transforms. IEEE Proc Commun Speech Vis 140(1):67–71
12. McClellan J, Chan D (1997) A 2-d fir filter structure derived from the chebyshev recursion. IEEE Trans Circ Syst 24(7):372–378
13. Kingsbury N (2002) Complex wavelets for shift invariant analysis and filtering of signals. J Appl Comput Harmonic Anal 10(3):234–253
14. Proenca H, Alexandre L. UBIRIS: a noisy iris image database. www.iris.di.ubi.pt
15. Multimedia University (2004) MMU iris image database. http://pesona.mmu.edu.my/ccteo
16. CASIA iris image database. http://www.sinobiometrics.com/casiairis.htm
17. IITD iris image database. http://web.iitd.ac.in
18. Daugman JG (1993) High confidence visual recognition of persons by a test of statistical independence. IEEE Trans Pattern Anal Mach Intell 25(11):1148–1161
19. Monro DM, Rakshit S, Zhang D (2007) Dct based iris recognition. IEEE Trans Pattern Anal Mach Intell 29(4):586–595
20. Sun Z, Tan T (2009) Ordinal measures for iris recognition. IEEE Trans Pattern Anal Mach Intell 31(12):2211–2226
21. Dong W, Tan T, Sun Z (2010) Iris matching based on personalized weight map. IEEE Trans Pattern Anal Mach Intell 99(1):1–14
22. Rahulkar AD, Holambe RS (2012) Partial iris feature extraction and recognition based on a new combined directional and rotated directional wavelet filter banks. Neurocomputing 1:12–23

Chapter 4
Iris Representation by Combined Hybrid Directional Wavelet Filter-Banks

Abstract This chapter addresses the issue in the design of DWFB and extends the proposed class of THFB in order to design hybrid finer directional wavelet filter bank. The iris feature extraction algorithm using proposed filter bank and post-classifier have been presented in this chapter.

Keywords Fan shaped filter bank · Hybrid directional wavelet filter bank · Iris recognition · Rotated directional filter bank

4.1 Introduction

In Chap. 2, a new class of THFB is introduced to extract iris feature. However, this method is capable to extract iris features in three directions only (horizontal, vertical, and mixed diagonal). In the Chap. 3, the cascade directional wavelet FB structure is suggested to extract the iris features in twelve different directions. In this, we have obtained the directional wavelet extension with the help of McClellan transformation on 1-D BWFB coefficients which are designed by the factorization of a HBP. However, this FB has the following limitations:

1. Ripples in the passband of the FSFB and CSFB.
2. Transition width is not narrow.
3. Computational complexity of the McClellan transformation is higher than the separable mapping.

In order to address these issues, this chapter generalized Ansari's method [1] to 2-D FSFB design using a structural approach based on three generalized HBPs. A three step ladder structure is used to design a new class of triplet halfband fan shaped filter bank (THFSFB) to provide more symmetry between analysis and synthesis filters than the conventional two-step ladder structure. The proposed class of FB provides desirable number of vanishing moments and more flexibility in the design. This THFSFB is sampled to obtain a pair of CSFB. The designed FB converts the wavelet basis functions to a set of directional basis elements by employing the combination of the designed class of 1-D THFB, THFSFB, and CSFB.

A. D. Rahulkar and R. S. Holambe, *Iris Image Recognition*,   59
SpringerBriefs in Signal Processing, DOI: 10.1007/978-3-319-06767-4_4,
© The Author(s) 2014

4.2 Review of the Related Filter Banks

Ansari et al. [1] described a structure based on three halfband filters. In their work, standard Lagrange interpolation formula is used to obtain the coefficients of these halfband filters. The objective of this structure is to overcome the inherent limitations of the structure based on two halfband filters (two-step ladder structure) [2]. Another related work is found in Eslami and Radha [3], where a new family of non-redundant geometrical image transform is proposed using CDF-9/7 filters and modified DFBs. These DFBs are designed by the triplet of half-band filters [1] and demonstrated the potential of proposed transform in non-linear approximation, image coding, and denoising applications. The objective of Lu and Do [4] was to provide the WT with finer directionality. In their work, CSFB is designed based on the parameterizations of the polyphase matrices. This form of parameterizations is proposed by Phoong et al. [2] which has some restrictions (see [1]). A number of approaches are proposed in the literature to provide more directional information (e.g. directionlets [5], hybrid wavelets and DBFs [3], multidimensional FB using triplet of Neville filters [6], finer directional wavelet filters [4], DFB [7], contourlets [8], curvelets [9], DT-CWT [10], RCWFs [11] etc.). However, most of the approaches have high computational complexity with redundancy. In this chapter, first 2-D separable wavelets are obtained by the proposed class of THFB (discussed in Chap. 2). With this, three directional filters are obtained (horizontal, vertical, and mixed diagonal). Secondly, a new class of THFSFB is designed using three generalized HBPs. This pair of THFSFB is passed through quincunx sampling matrix to obtain a pair of CSFB. This CSFB is applied separately on each of these three directional filters to obtain six hybrid directional filer bank (HDWFB). These six filters are rotated by angle of 45° to obtain six different rotated hybrid directional wavelet filter bank (RHDWFB). Further, these HDWFB and RHDWFB are combined together to obtain combined hybrid directional wavelet filter bank (CHDWFB) that provides twelve directional information.

4.3 Design of Combined Hybrid Directional Wavelet FB

4.3.1 Construction of 2-D Separable Filter Bank

The three detail and one approximation sub-bands are obtained as follows:

$$
\begin{aligned}
L_T(z) &= H_{0T}^{1d}(z_1) \times H_{0T}^{1d}(z_2) \\
H_T(z) &= H_{0T}^{1d}(z_1) \times H_{1T}^{1d}(z_2) \\
V_T(z) &= H_{1T}^{1d}(z_1) \times H_{0T}^{1d}(z_2) \\
D_T(z) &= H_{1T}^{1d}(z_1) \times H_{1T}^{1d}(z_2)
\end{aligned}
\tag{4.1}
$$

where $H_{0T}^{1d}(z_1)$ and $H_{1T}^{1d}(z_1)$ are the designed 1-D THFB LPF and HPF respectively. The design of $H_{0T}^{1d}(z_1)$ and $H_{1T}^{1d}(z_1)$ is discussed in Sect. 2.3.

The one-level decomposition results in one approximation sub-band (L_T), Horizontal (H_T), Vertical (V_T), and Diagonal (D_T) sub-bands which corresponds to LL, LH, HL and HH sub-bands respectively.

4.3.2 Design of the Triplet Halfband Fan Shaped Filter Bank

It is observed that HH sub-band of wavelets contains mixed two diagonal orientations $(+45°$ and $-45°)$. Consequently, separable wavelet FBs fail to separate these orientations. Also, separable wavelet FBs provide only three-directional information. In order to address these issues, we have constructed directional wavelet extension by FSFB with McClellan transformation in Chap. 3. However, this FSFB contains ripples in the passband of the FSFBs and consequently in CSFBs, does not provide narrow transition width of the FSFBs, and computational complexity of the McClellan transformation is higher than the separable mapping. Hence, we generalized Ansari's method [1] to a 2-D FSFB design using a structural approach based on the generalized HBPs. First, three 1-D HBPs are obtained by imposing the zeros at $z = -1$. On each of these three HBPs, 1-D to 2-D mapping suggested by Phoong et al. [2] is used whose complexity is that of separable FB and growing linearly with 1-D filter order. These mapped HBPs are used in three step ladder structure to obtain a new class of Triplet Halfband Fan Shaped Filter Bank (THFSFB). This scheme provides two degrees of freedom (free parameters of three HBPs and three step lifting scheme parameter p) leads to give more flexibility in the filter design.

First, general HBP $P(z)$ of an order K (expressed in coefficients a_k) which offers $(K/2) + 1$ degrees of freedom to impose the VMs is considered. From this polynomial, three HBPs $P_1(z)$, $P_2(z)$, and $P_3(z)$ are obtained by imposing M zeros at $z = -1$, where $M < (K/2)+1$. With this, desired number of independent parameters a_k (degree of freedom) are obtained without imposing maximum flatness constraint. These three HBPs can be expressed as follows:

$$P_i(z) = (z^{-1} + 1)^{M_i} R_i(z), \quad i \in 1, 2, 3. \tag{4.2}$$

where M_i is the number of zeros at $z = -1$ for ith polynomial and the remainder term $R_i(z)$ is given by the following equation:

$$R_i(z) = a_0 + c_1 z^{-1} + c_2 z^{-2} + \cdots + a_0 z^{K-M_i} \tag{4.3}$$

where c_j are the constants which can be expressed as functions of a single parameter a_0. Thus, three remainder polynomials $R_1(z)$, $R_2(z)$, and $R_3(z)$ are obtained by imposing M_1, M_2, and M_3 (where $M = M_1 + M_2 + M_3 = \frac{K}{2}$ can be a choice) zeros on $P(z)$ (of order K). It may be noted that the remainder polynomials can also be expressed into any desired number of independent or free parameters $(a_0, a_2, a_4, \ldots)$. Expressing the c_j (remainder polynomial) with more number of a_k provides better flexibility at the cost of computational complexity. With this, these three HBPs provide one degree of freedom (independent parameters) by which

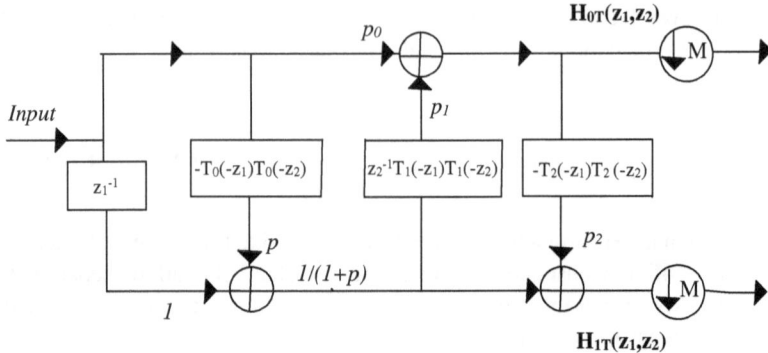

Fig. 4.1 Analysis bank of the triplet FSFB

flexibility in frequency responses can be achieved. The required class of three kernels for THFSFB is obtained by the following equations:

$$
\begin{aligned}
T_0(z) &= z^{K/2} P_1(z) - 1; \\
T_1(z) &= z^{K/2} P_2(z) - 1; \\
T_2(z) &= z^{K/2} P_3(z) - 1.
\end{aligned}
\tag{4.4}
$$

4.3.3   1-D to 2-D Mapping

In this subsection, the above 1-D framework generalized into 2-D THFSFB. It is known that the desired passband supports of the filters depend not only on the lattice but also on the choice of downsampling matrix/dilation matrix M [2]. The downsampling matrix used in this chapter is the quincunx sampling matrix as:

$$
M = \begin{bmatrix} 1 & 1 \\ 1 & -1 \end{bmatrix}
$$

The following transformation is used on the polyphase components of the three HBPs $T_0(z)$, $T_1(z)$, and $T_2(z)$:

1. Replace all the 1-D HBPs $T_i(z)$ with separable 2-D HBPs as $T_i(-z_1 z_2)$ $T_i(-z_1 z_2^{-1})$, where $i = 0, 1,$ and 2.
2. Replace the 1-D delay z^{-1} with 2-D delay as $z_1^{-1} z_2^{-1}$.

Under this transformation, the implementation of 2-D PR analysis FB is shown in Fig. 4.1. By using the noble identities, the analysis filters expressed as:

$$
H_{0T}(z_1, z_2) = \frac{1 + p}{2} + \frac{1}{2}(z_1^{K/2} z_2^{K/2} P_2(-z_1 z_2) P_2(-z_1 z_2^{-1}) - 1)
$$

$$
(1 + p(1 - z_1^{K/2} z_2^{K/2} P_1(-z_1 z_2) P_1(-z_1 z_2^{-1}))),
\tag{4.5}
$$

$$H_{1T}(z_1, z_2) = \frac{1}{1+p} + (1 + p((1 + p(1 - z_1^{K/2}z_2^{K/2}P_2(-z_1z_2)P_2(-z_1z_2^{-1}))))$$
$$-(\frac{1-p}{1+p}(z_1^{K/2}z_2^{K/2}P_3(-z_1z_2)P_3(-z_1z_2^{-1}) - 1)H_{0T}(z_1, z_2)).$$

(4.6)

In this work, six order generalized symmetric HBP is used to obtain three HBPs required for the construction of THFSFB. The value of $a_0 = -0.06245$ and $p = 0.414$ are selected to obtain the desired frequency response. Thus, the proposed design offers more flexibility in the design of filters using two degrees of freedom (a_k and p). The resulting lengths of the analysis LPF $H_{0T}(z_1, z_2)$ is ($N_1 + N_2 - 1 \times N_1 + N_2 - 1$) and analysis HPFs $H_{1T}(z)$ is ($N_1 + N_2 + N_3 - 2 \times N_1 + N_2 + N_3 - 2$), where N_1, N_2, and N_3 are the lengths of $T_0(z)$, $T_1(z)$, and $T_2(z)$ respectively. Figure 4.2 depicts the comparison of frequency responses of the FSFB obtained using 11/9 FB and McClellan transformation (discussed in Chap. 3), Phoong et al. [2], and the proposed class of THFSFB. It is observed that the proposed class of THFSFB yields smoother fan filters and thus introduces less ringing artifacts when employed in the DFBs.

This THFSFB $H_{0T}(z_1, z_2)$ and $H_{1T}(z_1, z_2)$ are sampled by the quincunx sampling matrix to obtain a pair of CSFB as:

$$C_{0T}(z_1, z_2) = H_{0T}(z_1z_2, z_1z_2^{-1})$$
$$C_{1T}(z_1, z_2) = H_{1T}(z_1z_2, z_1z_2^{-1})$$

(4.7)

In order to separate diagonal information and increase the directions of the proposed class of THFB, each HP sub-band is partitioned into two parts using a pair of CSFB. With this, six directional sub-bands are obtained which are approximately oriented in two mostly horizontal directions ($15°$ and $165°$), two mostly vertical directions ($75°$ and $105°$), and two separate diagonal directions ($45°$ and $135°$). It is observed that DT-CWT [10] also provides the same directional information. However it is four times redundant and used a different FB structure compared to wavelets. The proposed FB is derived by employing 1-D THFB, 1-D to 2-D mapping, and three step ladder structure, hence called as hybrid directional wavelet filter bank (HDWFB). RHDWFB is obtained by rotating HDWFB with an angle of $45°$. It is observed that HDWFB gives the information approximately oriented in $\theta_1 = (15°, 45°, 75°, 105°, 135°, 165°)$ and RHDWFB provides the information approximately oriented in $\theta_2 = (0°, 30°, 60°, 90°, 120°, 150°)$. The directional information of HDWFB and RHDWFB are combined together (CHDWFB) to obtain twelve-directional information as $\theta = (0°, 15°, 30°, 45°, 60°, 75°, 90°, 105°, 120°, 135°, 150°, 165°)$. Since, each individual filter of THFB is critically sampled; the overall FB is also critically sampled. The same set of directions is obtained by RCWFs [11]. However, RCWFB is obtained using known complex wavelet filter coefficients, which is two times redundant than the proposed HDWFBs.

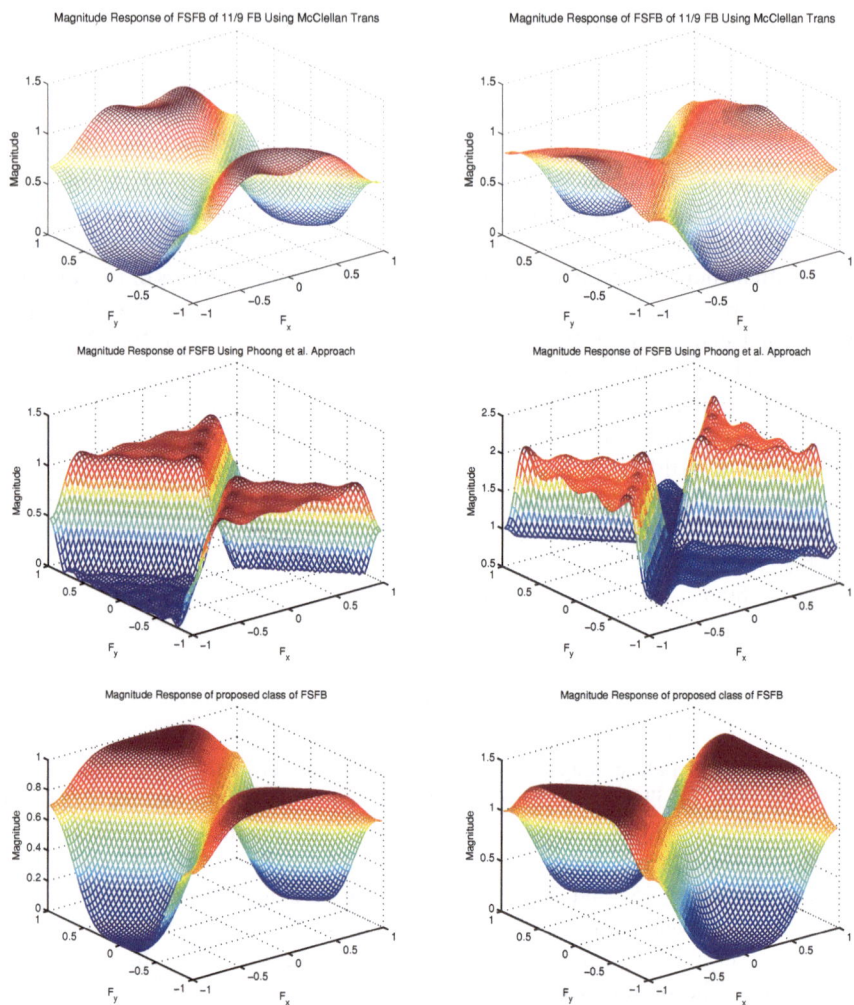

Fig. 4.2 *First row* Fan filter pair using McClellan transformation. *Second row* Fan filter pair using two-step ladder structure. *Third row* The proposed fan filter pair using three-step ladder structure

4.4 Iris Feature Extraction Using CHDWFB

The inner half iris region is divided into six sub-images and selected only four regions for further processing. The CHDWFB is applied separately on each of the four selected sub-blocks (sub-images). The feature vector for each sub-image is derived using the directional energies of the CHDWFB sub-bands. The four distance scores are obtained and *k-out-of-n:A* scheme is used to improve the recognition performance. The block diagram of the iris recognition scheme is shown in Fig. 4.3.

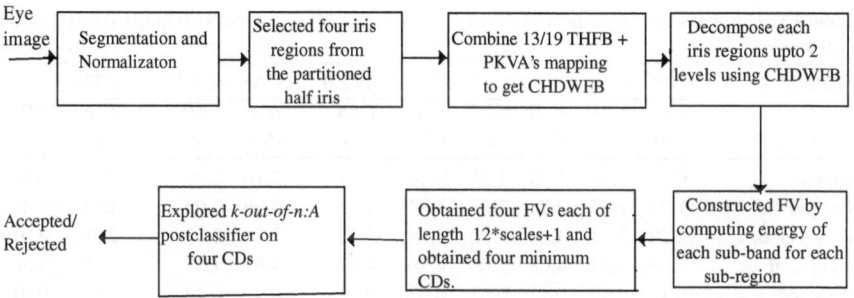

Fig. 4.3 Block diagram of proposed iris recognition algorithm

The preprocessing involves localization and normalization of iris image as mentioned in Sect. 2.4.1.

Since region closer to the pupil provides more useful iris information, the upper half iris from the entire normalized iris image is selected. This partial iris is divided into six sub-regions and selected only four sub-regions (Region: 1, 3, 4, and 6) as shown in Fig. 2.6d so as to minimize the effect of artifacts during iris recognition. The eyelids/eyelashes occlusions are minimum in these selected regions as compared to other two regions (Region: 2 and 5). The CHDWFB is applied on each of these four regions to extract twelve-directional iris texture. Each sub-band in CHDWFB represents iris texture characteristics in a particular scale and direction. The normalized directional energy is computed by L_1 norm from each sub-band of the CHDWFB at scale s and direction d. The feature vector is derived by concatenating the features at different scales and orientations. The derived feature vectors of each region are stored in the database as reference (enrollment process). The test iris pattern is classified on the basis of minimum CD between test iris FV and that of the FV of same person stored in database. In order to improve the performance, fusion at the decision level is incorporated using *k-out-of-n:A* scheme. Through this process, the final fused FARs and FRRs are obtained.

4.5 Experimental Results

This section evaluates the proposed approach using UBIRIS [12], MMU1 [13], CASIA-IrisV2.0 (device1) [14], CASIA-IrisV3-Interval [14], and IITD [15] databases. The proposed approach (combination of CHDWFB and *k-out-of-n:A*) has been compared with four successful existing iris recognition algorithms (Daugman [16], Monro et al. [17], Sun and Tan [18] and Dong et al. [19]). In order to assess the recognition accuracy of the postclassifier, the performance of proposed technique (CHDWFB + *k-out-of-n:A*) is compared for four different values of k with $n = 4$. The same set of experimentation is carried out as discussed in Chap. 3 to compute final fused FARs and FRRs. The performance of the proposed method has

Table 4.1 Comparison of scheme (CHDWFB + *k-out-of-n:A* (*k* = 2)) with existing iris recognition systems

Algorithms	UBIRIS		MMU1		CAS-IrisV2.0		CAS-IrisV3.0		IITD	
	FAR (%)	FRR (%)	FAR (%)	FRR (%)	FAR (%)	FRR (%)	FAR (%)	FRR (%)	FAR (%)	FRR (%)
Daugman	0.85	0.98	1.35	1.51	0.56	0.67	2.10	2.36	0.46	0.52
Sun and Tan	1.20	1.33	2.72	3.33	1.84	2.13	2.96	3.02	1.35	1.45
Monro et al.	2.29	3.11	4.64	5.56	3.59	4.27	5.16	5.11	2.98	3.01
Dong et al.	2.95	3.28	4.68	5.00	2.86	4.15	4.67	4.89	3.23	3.80
Proposed (k = 2)	0.12	0.17	0.52	0.65	0.074	0.083	0.71	0.79	0.011	0.013

Table 4.2 Comparison of scheme(CHDWFB + *k-out-of-n:A*) with different values of k

CHDWFB + k-out-of-n:A	UBIRIS		MMU1		CAS-IrisV2.0		CAS-IrisV3.0		IITD	
	FAR (%)	FRR (%)	FAR (%)	FRR (%)	FAR (%)	FRR (%)	FAR (%)	FRR (%)	FAR (%)	FRR (%)
k = 1	0.61	0.48	0.99	1.02	0.47	0.50	1.14	1.35	0.11	0.13
k = 2	0.12	0.17	0.52	0.65	0.074	0.083	0.71	0.79	0.011	0.013
k = 3	0.12	0.12	0.96	1.07	0.12	0.12	1.78	1.96	0.09	0.014
k = 4	2.19	2.30	6.20	6.43	1.79	1.89	12.47	12.72	1.99	2.06

been compared with four recent existing iris recognition algorithms. These include Daugman [16], Monro et al. [17], Sun and Tan [18], and Dong et al. [19]. These algorithms are implemented and tested on the same set of normalized iris images (without detection of eyelids, eyelashes, inaccurate localization, illumination variation, motion blur etc.) in order to have a fair comparison. A brief description of these existing iris recognition algorithms is given in Sect. 2.5. The experimental results have been presented in Table 4.1. From Table 4.1, it is observed that the proposed method yields superior performance against existing iris recognition methods. This is because the proposed scheme works on each region of iris independently, so artifacts can only affect the corresponding region and not the entire iris signature. The transformation on iris partitioned sub-regions does not corrupt the good iris region by combining them with artifacts (segmentation error, eyelids/eyelashes occluded regions, etc.). Thus, this approach achieves robustness under intra-class variations. Figure 4.4a–e show the comparison of the proposed scheme with existing iris recognition algorithms in the form of ROC curves on UBIRIS, MMU1, CASIA-IrisV2.0 (device1), CASIA-IrisV3-Inerval, and IITD databases respectively. The influence of *k-out-of-n:A* post-classifier on the recognition performance of the proposed method was tested for different values of k. The results are shown in Table 4.2.

It is observed from Table 4.2 that the recognition performance of *4-out-of-4:A* is very poor. It is because this post-classifier has been selected all the four iris regions for the recognition which affect the performance due to the artifacts present on the normalized iris image.

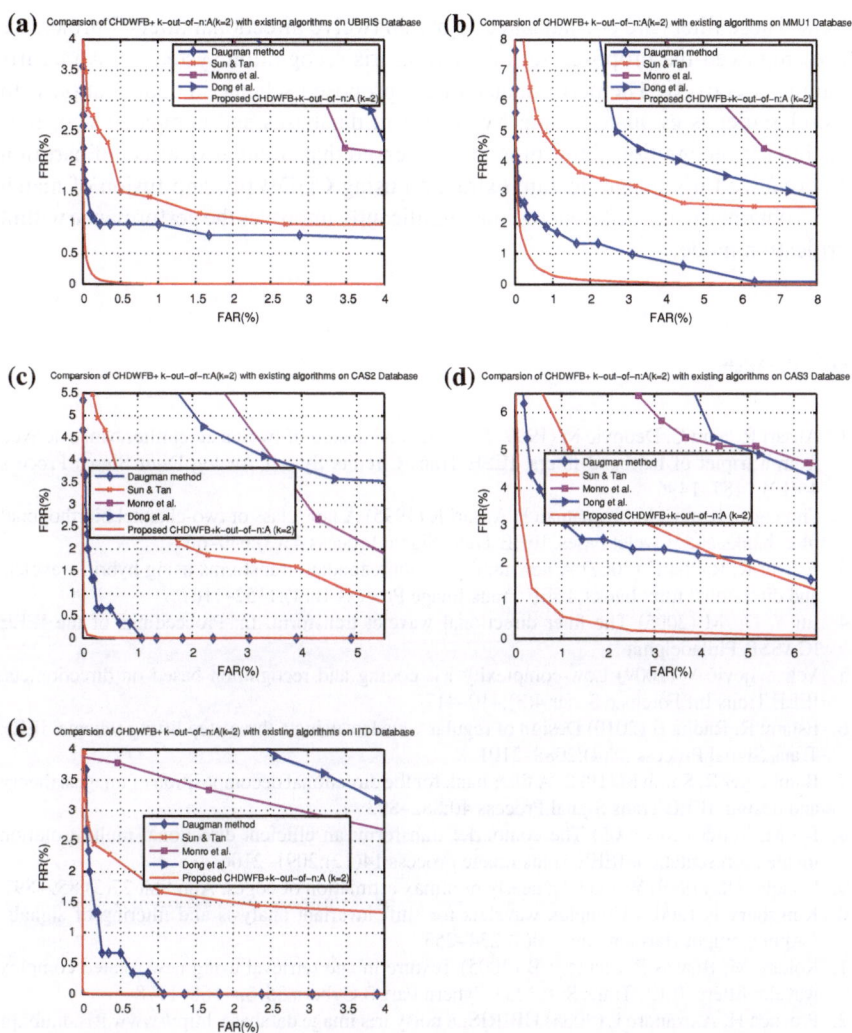

Fig. 4.4 Comparison of proposed algorithm with existing iris recognition algorithms using **a** UBIRIS **b** MMU1 **c** CASIA-IrisV2.0 **d** CASIA-IrisV3.0 **e** IITD iris databases

4.6 Summary

This chapter proposed a design method of 2-D THFSFB with desired number of vanishing moments using a three-step ladder structure. The proposed approach provides two degrees of freedom that achieve more flexibility in the filter design. Using this pair of THFSFB, checkerboard shaped filters are obtained that are applied on the detail sub-bands of the 2-D wavelets to obtain six finer directional information. These six filters are further rotated by angle of 45° to obtain six different directional filters. These filters are concatenated to obtain twelve directional filters. Further, we have employed the proposed set of filters in iris recognition system to extract iris features effectively. The performance of the proposed scheme in combination with post-classifier is evaluated using five different databases and compared with four existing iris recognition algorithms. It is observed that cumulative effect of partition of normalized iris image, feature extraction using CHDWFB, and fusion of match scores obtained using different regions significantly improves the performance within verification mode.

References

1. Ansari R, Kim C, Dedovic M (1999) Structure and design of two-channel filter banks derived from a triplet of halfband filters. IEEE Trans Circuits Syst II Analog Digit Signal Process 46(12):1487–1496
2. Phoong S, Kim C, Vaidyanathan P, Ansari R (1995) A new class of two-channel biorthogonal filter banks and wavelet bases. IEEE Trans Signal Process 43(3):649–665
3. Eslami R, Radha H (2007) A new family of nonredundant transforms using hybrid wavelets and directional filter banks. IEEE Trans Image Process 16(4):1152–1167
4. Lu Y, Do M (2005) The finer directional wavelet transform. In: Proceedings of the IEEE ICASSP, Philadelphia
5. Velisavljević V (2009) Low-complexity iris coding and recognition based on directionlets. IEEE Trans Inf Forensic Secur 4(3):410–417
6. Eslami R, Radha H (2010) Design of regular wavelets using a three-step lifting scheme. IEEE Trans Signal Process 58(4):2088–2101
7. Bamberger R, Smith M (1992) A filter bank for the directional decomposition of images: theory and design. IEEE Trans Signal Process 40:882–893
8. Do M, Vetterli M (2005) The contourlet transform: an efficient directional multiresolution image representation. IEEE Trans Image Process 14(12):2091–2106
9. Donoho DL (1999) Wedgelets: nearly minimax estimation of edges. Ann Stat 27(3):859–897
10. Kingsbury N (2002) Complex wavelets for shift invariant analysis and filtering of signals. J Appl Comput Harmon Anal 10(3):234–253
11. Kokare M, Biswas P, Chatterji B (2005) Texture image retrieval using new rotated complex wavelet fitlers. IEEE Trans Syst Man Cybern Part B Cybern 35(6):1168–1178
12. Proenca H, Alexandre L (2005) UBIRIS: a noisy iris image database. http://www.iris.di.ubi.pt
13. Multimedia University (2004) MMU iris image database. http://pesona.mmu.edu.my/ccteo
14. CASIA iris image database. http://www.sinobiometrics.com/casiairis.htm
15. IITD iris image database. http://web.iitd.ac.in
16. Daugman JG (1993) High confidence visual recognition of persons by a test of statistical independence. IEEE Trans Pattern Anal Mach Intell 25(11):1148–1161
17. Monro DM, Rakshit S, Zhang D (2007) DCT based iris recognition. IEEE Trans Pattern Anal Mach Intell 29(4):586–595
18. Sun Z, Tan T (2009) Ordinal measures for iris recognition. IEEE Trans Pattern Anal Mach Intell 31(12):2211–2226
19. Dong W, Tan T, Sun Z (2010) Iris matching based on personalized weight map. IEEE Trans Pattern Anal Mach Intell 99(1):1–14

Chapter 5
Ordinal Measures Based on Directional Ordinal Wavelet Filters

Abstract This chapter presents the design of the new class of triplet half-band checkerboard shaped filter bank (THCSFB) to solve the issue in the design of proposed non-separable FBs. This chapter also describes the directional ordinal measures (DOMs) for iris image representation based on THCSFB.

Keywords Checkerboard shaped filter bank · Iris recognition · Ordinal measures · Post-classifier · THFB

5.1 Introduction

Chapter 2 introduced a new class of THFB for iris feature extraction. However, this method is capable to extract iris features in three directions only (horizontal, vertical, and mixed diagonal). Chapters 3 and 4 proposed the directional wavelet FB structures to extract the iris features in twelve different directions. However, these structures need the cascade structure (combination of FSFB and CSFB) to improve the directionality of the WT. The approximate rotation invariance has been achieved by unwrapping the iris ring at five different initial angles. This method (presented in Chaps. 2–4) requires to store some additional iris FVs for different initial angles that leads to increase the memory requirement and computational complexity. It is also well accepted that the most challenging aspect in iris feature representation is to achieve the sensitivity to interclass differences and at the same time to maintain robustness against intraclass variations [1]. In order to answer these issues, this chapter:

1. introduces a directional ordinal measures (DOMs) scheme using a directional extension for wavelets based on the new class of THCSFB, and
2. uses the *k-out-of-n:A* postclassifier to handle possible artifacts especially segmentation error (inaccurate detection of inner and outer boundaries of iris), occlusion of eyelids/eyelashes, reflection on iris, non-linear deformation etc.

A. D. Rahulkar and R. S. Holambe, *Iris Image Recognition*,
SpringerBriefs in Signal Processing, DOI: 10.1007/978-3-319-06767-4_5,
© The Author(s) 2014

5.2 Review of the Related Filter Banks

Most of the iris feature extraction techniques in the literature used off-the-shelf wavelet basis to extract the iris features. However, many issues are still open in the field of iris feature-extraction and the choice of wavelet FB. The traditional way to obtain 2-D wavelets is to apply 1-D filters separately in horizontal and vertical directions. This provides only three directional components (horizontal, vertical, and mixed diagonal). It also suffers from poor diagonal orientation selectivity as 45° and 135° directions are combined into one subband in each resolution. There are a number of approaches available in the literature to provide more directional decompositions includes directionlets [2], DFB [3], contourlets [4], curvelets [5], complex wavelets [6], rotated complex wavelet filters [7], wedgelets [8], ridgelets [9], bandlets [10] etc. However, the WT is still very attractive in image processing applications for a number of reasons. First, the WT can be realized using a critically sampled FB that provides a non-redundant image decomposition. Secondly, there exist numerous algorithms and procedures utilizing wavelet for image processing applications, so one can benefit from these algorithms by cleverly adapting them to the transform family [11]. In this chapter, directional extension is provided to divide each HP sub-band of the wavelets into two finer directional sub-bands using a new class checkerboard support FB as the basic building block. The proposed FB provides the directional information approximately oriented in $\theta = (15°, 45°, 75°, 105°, 135°, 165°)$.

The objective of Lu and Do [12] was to provide the wavelet FB with finer directionality. In their work, a novel 1-D to 2-D mapping is suggested to design a CSFB based on the modified parameterizations of the polyphase matrices. This form of parameterizations is suggested by Phoong et al. [13] that is based on a pair of half-band filter bank (two-step ladder structure). The mapping (1-D to 2-D) proposed in [13] is only used to design 2-D diamond shaped and fan shaped filters. Eslami and Radha [11] constructed a new family of non-redundant hybrid geometrical image transforms using a THFB structure [14] and mapping given in [13]. This modified DFB shows good results in image coding and denoising at the cost of slow frequency roll-off. The same authors [15] generalized Ansari's method [14] to a multidimensional FB design using a structural approach based on Kovacevic method [16]. Tanaka et al. [17] presented multiresolution image representation using combined 2-D and 1-D directional filter banks based on the design given in [18]. Kim and Udpa [19] introduced a set of rotated wavelet filters (RWFs) using Haar wavelet coefficients to separate the diagonal information. However, this approach has limited directionality. Kokare et al. [7] constructed rotated complex wavelet filters (RCWFs) using known complex wavelet filter coefficients to overcome the limitations of wavelet filters. The work presented in [7] used the same approach as given in [6]. This set of RCWFs is two times redundant than wavelet filters.

5.3 Design of Triplet Halfband Checkerboard Shaped Filter Bank

In this chapter, Triplet Halfband Checkerboard Shaped Filter Bank (THCSFB) is designed using three generalized HBPs in order to tile the wavelet sub-bands into two parts.

5.3.1 Design of Triplet 1-D Framework from Generalized HBPs

The input image is first decomposed by the separable THFB to give four subbands LL, LH, HL, and HH for each level. The three detail subbands (LH, HL, and HH) suffer from limited directionality (different directional frequencies are gathered into one subband) which are unable to give effective multiresolution image representation. Hence, the new class of THCSFB is proposed to improve its directionality. First, general HBP $P(z)$ of an order K (expressed in coefficients a_k) which offers $(K/2)+1$ degrees of freedom to impose the VMs is considered. From this polynomial, three HBPs $P_1(z)$, $P_2(z)$, and $P_3(z)$ are obtained by imposing M zeros at $z = -1$, where $M < (K/2) + 1$. With this, desired number of independent parameters a_k (degree of freedom) are obtained without imposing maximum flatness constraint. These three HBPs can be expressed as follows:

$$P_i(z) = (z^{-1} + 1)^{M_i} R_i(z), \quad i = 1, 2, 3. \tag{5.1}$$

where M_i is the number of zeros at $z = -1$ for ith polynomial and the remainder term $R_i(z)$ is given by the following equation:

$$R_i(z) = a_0 + c_1 z^{-1} + c_2 z^{-2} + \cdots + a_0 z^{K-M_i} \tag{5.2}$$

where c_j are the constants which can be expressed as functions of a single parameter a_0. Thus, three remainder polynomials $R_1(z)$, $R_2(z)$, and $R_3(z)$ are obtained by imposing M_1, M_2, and M_3 (where $M = M_1 + M_2 + M_3 = \frac{K}{2}$ can be a choice) zeros on $P(z)$ (of order K). It may be noted that the remainder polynomials can also be expressed into any desired number of independent or free parameters $(a_0, a_2, a_4, \ldots)$. Expressing the c_j (remainder polynomial) with more number of a_k provides better flexibility at the cost of computational complexity. With this, these three HBPs provide one degree of freedom (independent parameters) by which flexibility in frequency responses can be achieved. The required class of three kernels for CSFB is obtained by following equation:

$$T_0(z) = z^{K/2} P_1(z) - 1;$$
$$T_1(z) = z^{K/2} P_2(z) - 1;$$
$$T_2(z) = z^{K/2} P_3(z) - 1.$$

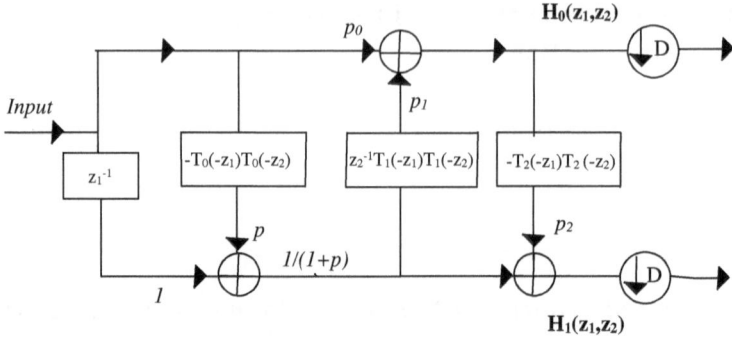

Fig. 5.1 Analysis bank of the THCSFB

5.3.2 1-D to 2-D Mapping

In this subsection, the above 1-D framework is generalized into 2-D THCSFB. It is known that the desired passband supports of the filters depend not only on the lattice but also on the choice of downsampling matrix/dilation matrix D. The downsampling matrix used in this chapter is the simple diagonal matrix as:

$$D = \begin{bmatrix} 2 & 0 \\ 0 & 2 \end{bmatrix}$$

The following transformation is used on the polyphase components of the three HBPs $T_0(z)$, $T_1(z)$, and $T_2(z)$:

1. Replace all the 1-D HBPs $T_i(z)$ with separable 2-D HBPs as $T_i(-z_1)T_i(-z_2)$, where $i = 0, 1, 2$.
2. Replace the 1-D delay z^{-1} with 2-D delay as $z_1^{-1}z_2^{-1}$.

Under this transformation, the implementation of 2-D PR analysis FB is shown in Fig. 5.1. By using the noble identities, the analysis and synthesis filters expressed as:

$$H_0(z_1, z_2) = \frac{1+p}{2} + \frac{1}{2}T_1(-z_1)T_1(-z_2) - \frac{p}{2}T_0(-z_1)T_0(-z_2)T_1(-z_1)T_1(-z_2),$$

$$G_0(z_1, z_2) = \frac{1+pT_0(-z_1)T_0(-z_2)}{1+p} + \frac{1-p}{1+p}T_2(-z_1)T_2(-z_2)[\frac{1+p}{2} - \frac{1}{2}T_1(-z_1)T_1(-z_2)$$
$$(1 + pT_0(-z_1)T_0(-z_2))],$$

$$H_1(z_1, z_2) = -z_1^{-1}G_0(-z_1, z_2),$$

$$G_1(z_1, z) = z_1^{-1}H_0(-z_1, z_2).$$

The main differences between our method and Lu and Do [12] are:

1. Lu and Do [12] used traditional two-step lifting scheme, whereas we have used three step ladder structure to provide more flexibility in the design.

2. Lu and Do used Parks-McClellan transformation to obtain the required coefficients, whereas we have designed the three HBPs using generalized symmetric HBP by imposing zeros at $z = -1$ to obtain the required coefficients.
3. The proposed approach has two degrees of freedom (coefficients of HBPs and shape parameter p) than one in the existing design of Lu and Do [12].
4. Mapping proposed by Lu and Do required upsampling by two operation $(z_1^{-1}$ $\alpha(-z_1^2, z_2))$ to obtain the required support during the construction of CSFB, whereas we do not require this operation (inherent in the HBPs).

5.3.3 Properties of the 1-D to 2-D Mapping

The proposed THCSFB satisfies the following properties:

1. **Perfect reconstruction**: The 1-D to 2-D mapping preserved the perfect reconstruction property due to the implicit nature of ladder structure.
2. **Critically sampled**: As each individual component of the proposed directional extension (combination of wavelet highpass and THCSFB) is critically sampled, the overall system is also critically sampled.
3. **Linear phase**: The linear phase property of the analysis and synthesis filters is preserved by this 2-D mapping.
4. **Complexity**: Though the designed 2-D filers are nonseparable, their polyphase components are separable including the downsampling matrix D. Thus, the whole system can be implemented by 1-D operations only.

In this work, six order generalized symmetric HBP is used to obtain three HBPs required for the construction of THCSFB. The value of $a_0 = -0.06245$ and $p = 0.414$ are selected to obtain the desired frequency response. Thus, the proposed design offers more flexibility in the design of filters using two degrees of freedom (a_k and p). The resulting lengths of the analysis LPF $H_0(z_1, z_2)$ is $(N_1 + N_2 - 1 \times N_1 + N_2 - 1)$ and analysis HPFs $H_1(z)$ is $(N_1 + N_2 + N_3 - 2 \times N_1 + N_2 + N_3 - 2)$, where N_1, N_2, and N_3 are the lengths of $T_0(z)$, $T_1(z)$, and $T_2(z)$ respectively. In order to validate the design, the frequency responses of the proposed class of THCSFB are compared with the frequency responses obtained in [12] as shown in Fig. 5.2.

5.3.4 Directional Extension of Wavelet Filter Bank

The detail subbands (LH, HL, and HH) obtained by the separable THFB (given in Sect. 2.3) are further decomposed by the THCSFB. At the first level of construction, three detail sub-bands and one approximation sub-band (LH, HL, HH, and LL) are obtained by employing the proposed separable class of THFB. During the second level, a pair of THCSFB is applied on each detail sub-band using Eq. 5.3 as:

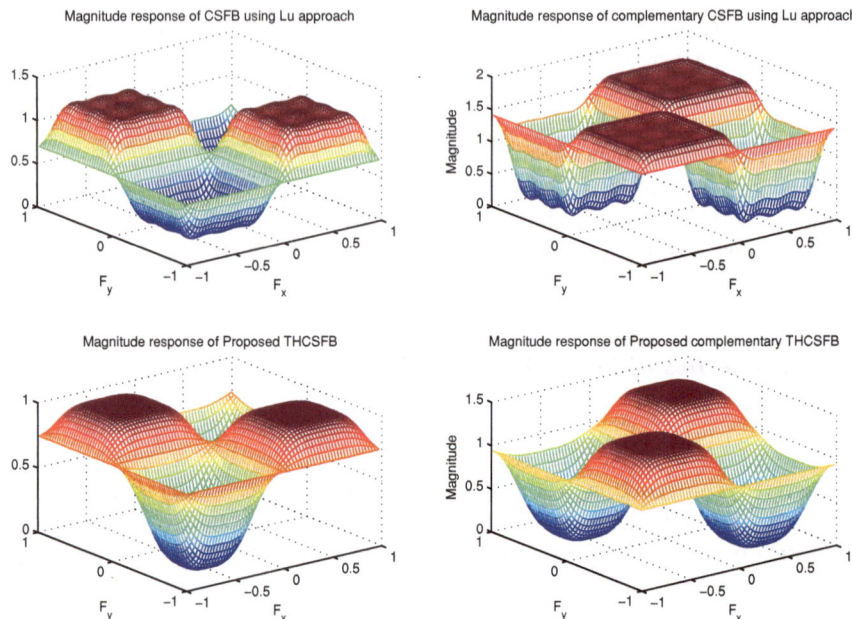

Fig. 5.2 *First row* Pair of checkerboard shaped FB using Lu and Do method. *Second row* Pair of the proposed class of THCSFB

$$
\begin{aligned}
H_{00}(z_1, z_2) &= H(z_1, z_2)H_0(z_1, z_2), \\
H_{01}(z_1, z_2) &= H(z_1, z_2)H_1(z_1, z_2), \\
H_{02}(z_1, z_2) &= V(z_1, z_2)H_0(z_1, z_2), \\
H_{03}(z_1, z_2) &= V(z_1, z_2)H_1(z_1, z_2), \\
H_{04}(z_1, z_2) &= D(z_1, z_2)H_0(z_1, z_2), \\
H_{05}(z_1, z_2) &= D(z_1, z_2)H_1(z_1, z_2).
\end{aligned}
\tag{5.3}
$$

where $H(z_1, z_2)$, $V(z_1, z_2)$, and $D(z_1, z_2)$ corresponds to horizontal, vertical, and diagonal sub-bands of the separable THFBs. With this, each detail sub-band of THFBs gives two-directional information using THCSFB. Thus, this structure gives two mostly horizontal directions (15° and 165°), two mostly vertical directions (75° and 105°), and two separate diagonal directions (45° and 135°). The construction of this structure is shown in Fig. 5.3.

The synthetically generated image is decomposed using proposed structure up to one level and shown in Fig. 5.4 in order to visualize six-directional information. It is observed that proposed structure provides separate sub-bands for features approximately oriented in directions $\theta = (15°, 45°, 75°, 105°, 135°, 165°)$. The same set of directions have been obtained by dual-tree complex wavelet transform in [6], however it is four times redundant and used a different structure for design

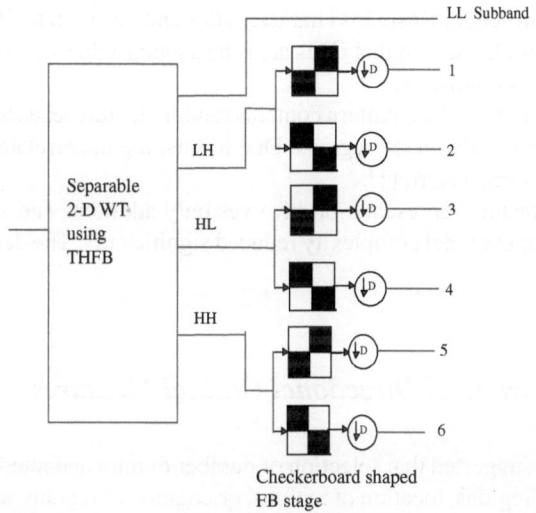

Fig. 5.3 Complete structure of the directional extension using THCSFB

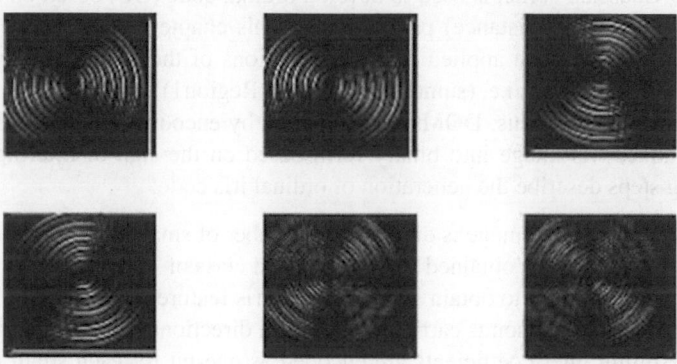

Fig. 5.4 One-level decomposition of image using proposed class of non-separable FBs

5.4 Directional Ordinal Measures for Iris Recognition

5.4.1 Introduction to Ordinal Measures

Ordinal Measures (OMs) encode qualitative information of visual signal rather than its quantitative values [1]. It has been proved in [1] that OMs are capable to represent the distinctive and robust features of iris patterns. The properties of OMs desirable for accurate and robust iris recognition are as follows [1]:

1. The OMs represent intrinsic iris characteristics and invariant to illumination variations [1]. It is also shown that OMs are robust against dust on eyeglasses, partial
   occlusion, sensor noise etc.
2. It is well accepted that iris pattern contains random texture leads to sharp intensity
   variations across iris image regions. Due to this, the uncorrelated OMs provide
   uniqueness of iris pattern [1].
3. OMs based feature representation involves only additions and subtractions due
   to which computational complexity reduced significantly. The details of OMs are
   found in [1].

5.4.2 Construction of Directional Ordinal Measures

Sun and Tan [1] suggested that selection of number of intra and inter-region parameters i.e. shape of regions, location of regions, orientation of regions, average intensity
feature type, wavelet coefficients feature type, inter-region distance, spatial configu-
ration of regions etc. provide immense flexibility in order to design specific ordinal
iris feature extraction scheme. In their work, multi-lobe differential filter (MLDF)
using 2-D Gaussian kernel is used to develop ordinal code (OC) based on intralobe
(scale) and interlobe (distance) parameters. In this chapter, the proposed class of
six directional filters are applied on smaller regions of the normalized iris image
and ordinal comparisons i.e. (sum(Region2)-sum(Region1)) are carried out for each
directional filter. With this, DOMs are computed by encoding the small regions of
the normalized iris image into binary form based on the sign of filtering results.
Following steps describe the generation of ordinal iris code:

1. The normalized iris image is divided into number of small regions (say r).
2. Six directional filters obtained by the proposed class of THCSFB are applied on
   these r regions so as to obtain six directional iris features.
3. The ordinal comparison is carried out for each directional iris features.
4. The resulting filter coefficients are encoded as one-bit for each small region r.
   All these bits are combined together to derive a single iris OC.

5.5 Iris Feature Extraction Using Proposed DOMs

The block diagram of the proposed iris recognition algorithm is shown in Fig. 5.5.
In this work, inner half iris region has been divided into six sub-images and ordinal
iris code is derived from each sub-image by DOMs. The dissimilarity between two
ordinal iris codes has been computed by Hamming distance (HD) for each sub-
image. Six individual Hamming distances are obtained for six sub-images. These six
dissimilarity scores are fused using proposed *k-out-of-n:A* postclassifier (discussed
in Sect. 2.4.2) in order to develop more robust iris recognition system.

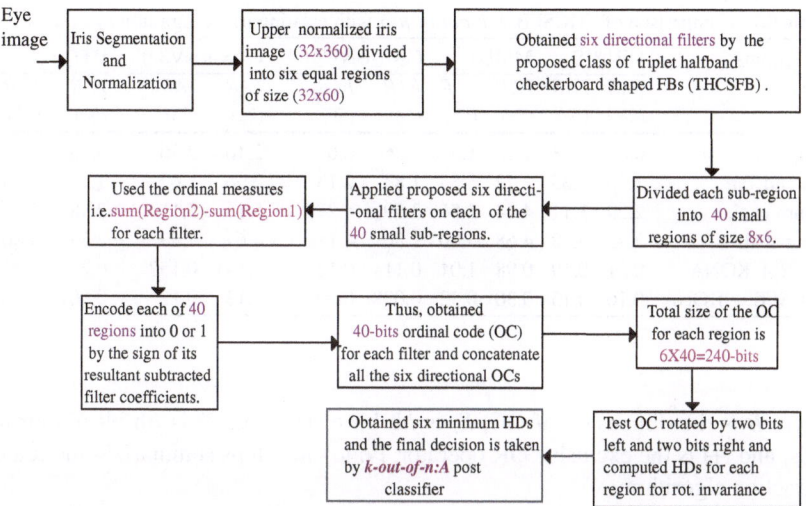

Fig. 5.5 Block diagram of the proposed directional ordinal based iris recognition

First, iris is localized using Daugman's IDO and normalized with the help DRSM of the fixed size [20]. The preprocessing steps used same as shown in Fig. 2.6a–c. Although some of the existing methods extract iris texture efficiently, their performance degrades significantly when the image quality is poor. Chen et al. [21] suggested that different regions of the iris have different qualities and local iris image regions with better quality have better classification capability. In multi-biometric recognition system, fusion of information extracted from classifiers provide better recognition performance as compared to single classifier [22]. Therefore, instead of recognizing the entire iris image, we have divided the iris image into multiple regions. Each iris sub-region recognized separately and fused the decision using *k-out-of-n:A* postclassifier. The upper half-iris (inner iris region) is preferred from the entire normalized iris image because (1) region closer to the pupil provides more discriminating iris information and (2) limbic boundary sometimes may not be segmented properly. Further, this half-iris is divided into six non-overlapping sub-images as shown in Fig. 2.6d to improve the performance of iris recognition system.

Each sub-image is further divided into a number of small regions and the proposed class of six directional filters are applied on these small regions so as to encode each small region into 1-bit. Thus, total six directional ordinal codes are combined together to obtain the single ordinal code (OC) for each sub-image of the normalized iris image. The total derived OC of each sub-image is stored in the database as reference (enrollment process). The test iris pattern is classified using Hamming distance. The Hamming distance is between test OC and that of the database is computed as

$$HD(X, Y) = \frac{1}{B} \sum_{i=1}^{B} X_i \bigoplus Y_i \qquad (5.4)$$

Table 5.1 Comparison of THCSFB + *k-out-of-n:A* with existing iris recognition systems

Algorithms	UBIRIS		MMU1		CAS-IrisV2.0		CAS-IrisV3.0		IITD	
k = 2	FAR	FAR	FAR	FAR	FAR	FAR	FAR	FAR	FAR	FAR
	(%)	(%)	(%)	(%)	(%)	(%)	(%)	(%)	(%)	(%)
Daugman	0.8	0.98	1.35	1.51	0.56	0.67	2.10	2.36	0.46	0.52
Sun and Tan	1.20	1.33	2.72	3.33	1.84	2.13	2.96	3.02	1.35	1.45
Monro et al.	2.29	3.11	4.64	5.56	3.59	4.27	5.16	5.11	2.98	3.01
Dong et al.	2.95	3.28	4.68	5.00	2.86	4.15	4.67	4.89	3.23	3.80
THFB + KONA	0.21	0.19	0.98	1.01	0.11	0.12	0.141	0.157	0.2	0.23
THCSFB + KONA	0.16	0.15	0.30	0.32	0.098	0.093	0.13	0.12	0.001	0.008

where B is the dimension of OC, X_i is ith bit of test OC, Y_i is ith bit of enrolled OCs, and $\oplus$ is the exclusive OR operator. Following steps summarizes the feature extraction algorithm:

1. Iris image is segmented using IDO.
2. The segmented iris is normalized using DRSM.
3. The upper half iris is selected from the entire normalized iris image.
4. This partial iris is divided into six sub-images for the verification.
5. Each sub-image is partitioned into number of small regions.
6. OC for each sub-image is derived by the procedure described in Sect. 5.5. Thus total six OCs have been derived for the verification.

In order to improve the performance of the proposed encoding scheme, fusion at the decision level is incorporated using *k-out-of-n:A* scheme. Through this process, the final fused FARs and FRRs are obtained.

5.6 Experimental Results

This section presents the experimental evaluation of the proposed approach using UBIRIS, MMU1, CASIA-Iris V2.0, CASIA-IrisV3.0, and IITD databases. The proposed approach is compared with four state-of-the-art iris recognition algorithms (Daugman [20], Monro et al. [23], Sun and Tan [1], and Dong et al. [24]) in order to assess the accuracy of the proposed method.

The approximate translation and scale invariance are achieved by localizing the iris from an eye image and normalizing it into fixed size. The registration process used one iris image per subject. To achieve rotation invariance, test OC is circularly shifted towards left and right in order to match the enrolled OCs. The minimum Hamming distance of the same class has been selected as the final score.

For the testing, remaining four iris images per class are used. In order to minimize the effect of intra-class variations and avail the discriminating iris information for the recognition, the upper half iris part from the original normalized iris image is

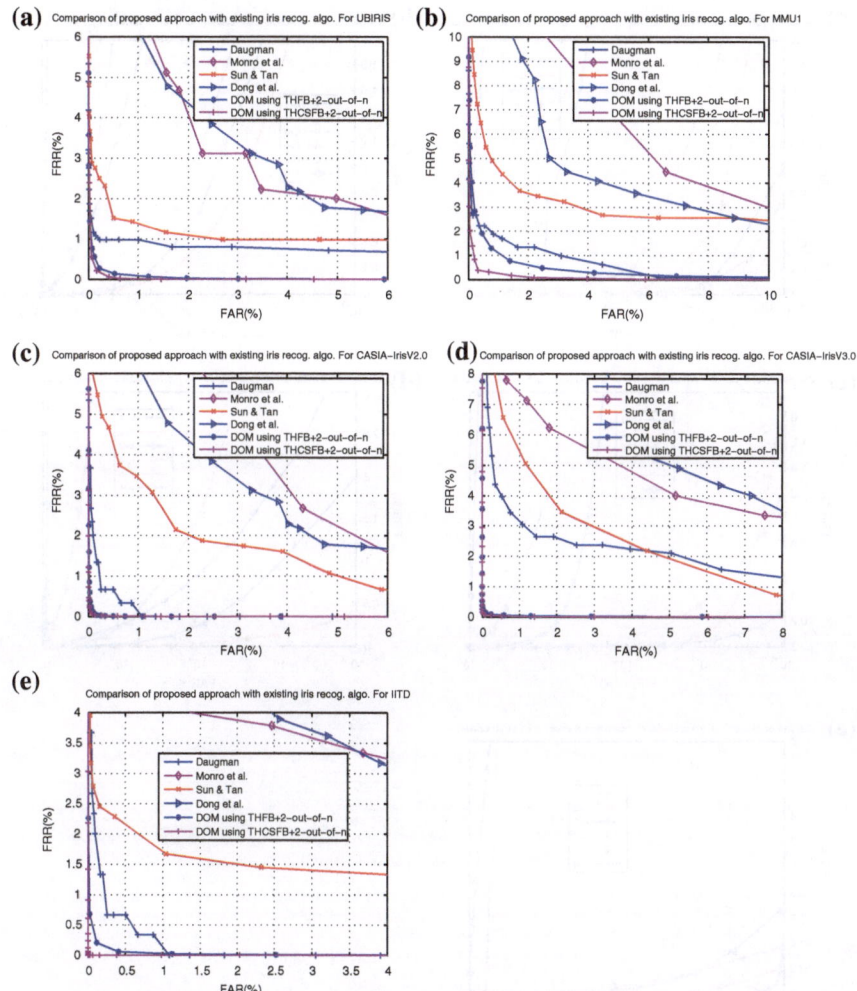

Fig. 5.6 Comparison of proposed algorithm with existing iris recognition algorithms using **a** UBIRIS **b** MMU1 **c** CASIA-IrisV2.0 **d** CASIA-IrisV3.0 **e** IITD iris databases

selected. The selected half iris is partitioned into six sub-images and used for the further processing. Each of these six sub-images is divided into 40 small regions of size 8×6 and the proposed class of six directional filters with one scale are applied on these regions to create six 40-bits iris OCs separately (size of feature vector = $6 \times 40 = 240$-bits). The artifacts present in iris images lead to reduction in accuracy. In order to reduce the error rate, *k-out-of-n:A* postclassifier is used on these six OCs. During training, a set of six independent ROCs (from selected six sub-images) is obtained from the corresponding inter-class and intra-class comparisons. These ROCs are fused by the postclassifier to get a single ROC. From this single

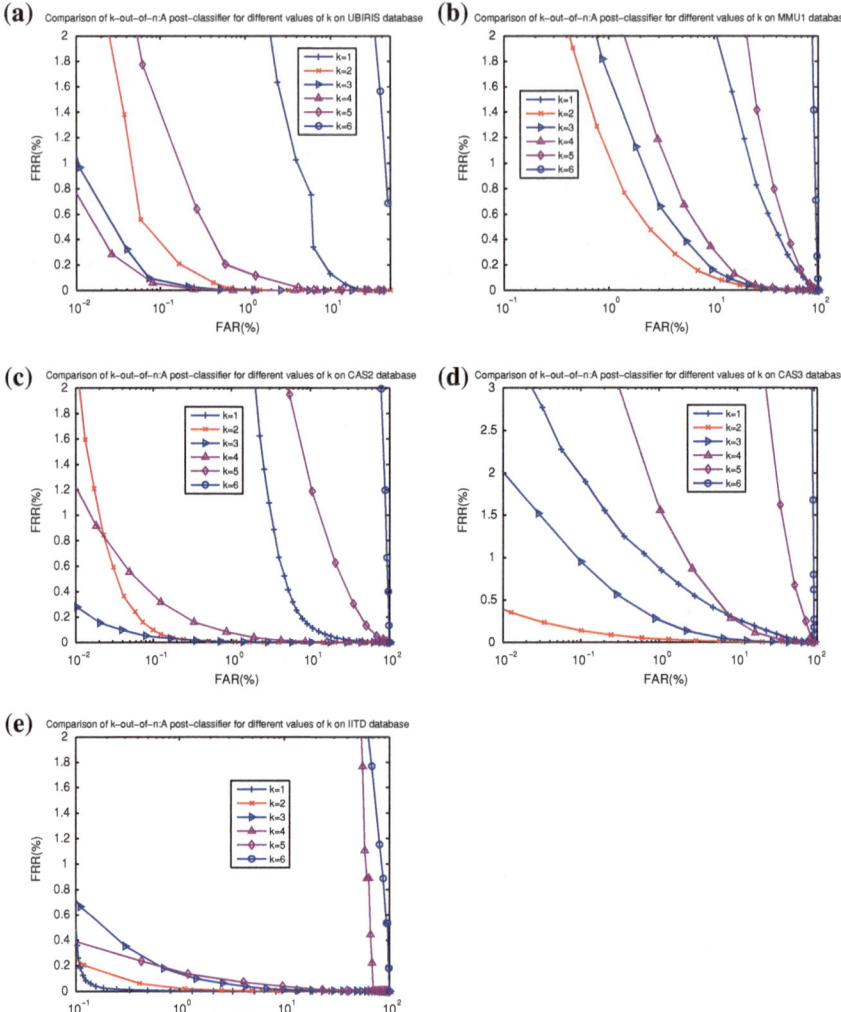

Fig. 5.7 Comparison of postclassifier using **a** UBIRIS **b** MMU1 **c** CASIA-IrisV2.0 **d** CASIA-IrisV3.0 **e** IITD iris databases

ROC, optimal operating point is decided by the possible combinations of FARs and FRRs. Using this optimal operating point, thresholds of corresponding regions are obtained. During the testing, final decision is taken by the postclassifier with the help of six HDs. The same set of experimentations has been carried out using proposed class of separable THFB (three directions). The performance of the proposed method with $k = 2$ has been compared with separable THFB and four existing well known iris recognition algorithms. These include Daugman [20], Monro et al. [23], Sun and Tan [1], and Dong et al. [24]. These algorithms are implemented and tested on the same set of normalized iris images (without detection of eyelids, eyelashes, inaccurate localization, illumination variation, motion blur etc.) in order to have a fair comparison. The experimental results have been presented in Table 5.1 It is observed from Table 5.1 that the proposed method yields superior performance against existing iris recognition methods. This is because the proposed scheme works on each selected region of iris independently, so artifacts can only affect the corresponding region and not the entire iris signature. The transformation on iris partitioned sub-regions does not corrupt the good iris region by combining them with artifacts. Thus, the proposed approach achieves the robustness to intra-class variations (especially occlusion of pupil on iris due to inaccurate pupil segmentation, occlusion of eyelids/eyelashes, illumination variation due to the use of OM etc.). Figures 5.6a–e show the comparison in the form of ROC curve of the proposed method with state-of-the-art algorithms on UBIRIS, MMU1, CASIA-IrisV2.0-device1, CASIA-IrisV3.0-interval, and IITD databases respectively. The performance of the proposed method is better than the existing iris recognition methods on all the five iris databases.

The influence of k-out-of-n:A post-classifier on the recognition performance of the proposed method was tested for different values of k. The performance of the proposed DOM based technique has been compared for six different values of k i.e. $k = 1, k = 2, k = 3, k = 4, k = 5, k = 6$ and $n = 6$. Figures 5.7a–e show the ROC curves of these methods on UBIRIS, MMU1, CASIA-Iris-V2-device1, CASIA-IrisV3.0-interval, and IITD iris databases. It is observed that 2-out-of-6:A (acceptance of any two regions out of 6-regions) and 3-out-of-6:A (acceptance of any three regions out of 6-regions) post-classifiers best suited for all the databases in order to improve the recognition performance of the proposed method. The recognition performance of 6-out-of-6:A is very poor. It is because this post-classifier has been selected all the six iris regions for the recognition which affect the performance due to the aforementioned artifacts present on the normalized iris.

5.7 Summary

In this chapter, a new directional iris ordinal code based on the new class of THCSFB is derived. The combined desirable properties of the proposed FB and OMs are utilized for the effective and efficient iris feature extraction. Also, flexible postclassifier is used on partitioned sub-images in order to reduce the FR. The developed method is robust against inaccurate partial-pupillary and limbic boundary segmentations, and occlusion of eyelids/eyelashes. Also, the method is invariant to shift, scale and rotation. The performance of the presented scheme has been evaluated using five different databases and compared with four recently developed iris recognition algorithms. It is observed that cumulative effect of partitioning of normalized iris image, feature extraction using DOM, and use of k-out-of-n:A postclassifier significantly reduces FRR.

References

1. Sun Z, Tan T (2009) Ordinal measures for iris recognition. IEEE Trans Pattern Anal Mach Intell 31(12):2211–2226
2. Velisavljević V (2009) Low-complexity iris coding and recognition based on directionlets. IEEE Trans Inf Forensic Secur 4(3):410–417
3. Bamberger R, Smith M (1992) A filter bank for the directional decomposition of images: theory and design. IEEE Trans Signal Process 40:882–893
4. Do M, Vetterli M (2005) The contourlet transform: an efficient directional multiresolution image representation. IEEE Trans Image Process 14(12):2091–2106
5. Candes E, Demanet L, Donoho D (2005) Fast discrete curvelet transform. Technical report, CalTech
6. Kingsbury N (2002) Complex wavelets for shift invariant analysis and filtering of signals. J Appl Comput Harmon Anal 10(3):234–253
7. Kokare M, Biswas P, Chatterji B (2005) Texture image retrieval using new rotated complex wavelet fitlers. IEEE Trans Syst Man Cybern B Cybern 35(6):1168–1178
8. Donoho DL (1999) Wedgelets: nearly minimax estimation of edges. Ann Stat 27(3):859–897
9. Candes E, Donoho DL (1999) Ridgelets: a key to higher dimensional intermittency? Philos Trans R Soc Lond 357:2495–2509
10. Pennec E, Mallat S (2005) Sparse geometric image representation with bandelets. IEEE Trans Image Process 14(4):423–438
11. Eslami R, Radha H (2007) A new family of nonredundant transforms using hybrid wavelets and directional filter banks. IEEE Trans Image Process 16(4):1152–1167
12. Lu Y, Do M (2005) The finer directional wavelet transform. In: Proceedings of the IEEE ICASSP, Philadelphia
13. Phoong S, Kim C, Vaidyanathan P, Ansari R (1995) A new class of two-channel biorthogonal filter banks and wavelet bases. IEEE Trans Signal Process 43(3):649–665
14. Ansari R, Kim C, Dedovic M (1999) Structure and design of two-channel filter banks derived from a triplet of halfband filters. IEEE Trans Circuits Syst II Analog Digit Signal Process 46(12):1487–1496
15. Eslami R, Radha H (2010) Design of regular wavelets using a three-step lifting scheme. IEEE Trans Signal Process 58(4):2088–2101
16. Kovacevic J, Sweldens W (2000) Wavelet families of increasing order in arbitrary dimensions. IEEE Trans Image Process 9(3):480–496
17. Tanaka Y, Ikehara M, Nguyen TQ (2009) Multiresolution image representation using combined 2-D and 1-D directional filter banks. IEEE Trans Image Process 2:466–480
18. Tay DBH, Kingsbury N (1993) Flexible design of multidimensional perfect reconstruction FIR 2-band filter-banks using transformation of variables. IEEE Trans Image Process 2:466–480
19. Kim N, Udpa S (2000) Texture classification using rotated wavelet filters. IEEE Trans Syst Man Cybern Part A Syst Hum 30(6):847–852
20. Daugman JG (1993) High confidence visual recognition of persons by a test of statistical independence. IEEE Trans Pattern Anal Mach Intell 25(11):1148–1161
21. Chen Y, Dass S, Jain A (2006) Localized iris image quality using 2-D wavelets. In Proceedings of international conference on biometrics, pp 373–381
22. Ross A, Jain A (2003) Information fusion in biometrics. Pattern Recogn Lett 24(13):2115–2125
23. Monro DM, Rakshit S, Zhang D (2007) DCT based iris recognition. IEEE Trans Pattern Anal Mach Intell 29(4):586–595
24. Dong W, Tan T, Sun Z (2010) Iris matching based on personalized weight map. IEEE Trans Pattern Anal Mach Intell 99(1):1–14

Appendix A

The extensive experiments were conducted on five datasets, namely UBIRIS [1], MMU1 [2], CASIA-IrisV2.0 (device1) [3], CASIA-IrisV3 (Interval) [3], and IITD [4] databases. The images have been captured with different instruments under varying conditions and different ethnicity. UBIRIS database consists of 1,877 iris images of 241 persons captured in two different sessions using Nikon E5700 camera. The images in this database have artifacts in the form of reflection, contrast, natural luminosity, focus and eyelids/eyelashes obstructions (leads to intra-class variations of iris). MMU1 database consists of eye images of 45 persons having 5 images of each eye. It includes total 450 images of 90 subjects. These images are captured with LG IrisAccess camera and contain severe obstructions by eyelids/eyelashes, specular reflection, non-linear deformation, low contrast and illumination changes. CASIA-Iris V2.0 database includes two subsets captured with two different devices (device1-OKI Irispass-h and device2-CASIA-Iris CamV2). Each subset includes 1,200 images from 60 classes. These images contain occlusion of eyelids, eyelashes, blur, illumination variation, slight shadow of eyelids/eyelashes etc. The CASIA-IrisV3-Interval database has the left ('L') and right ('R') irides of 249 subject and 396 eye images. The total number of images in this database is 2,655. These images are captured with self-developed iris sensor and contain eyelids/eyelashes occlusion, pupil dilation, head-tilt, specular reflection, slightly shadow of eyelids on iris etc. IIT Delhi (IITD) database is the first Indian database consists of total 1,120 iris images from 224 subjects. This database consists of low resolution images and these images are captured using JIRIS, JPC1000, and digital CMOS camera. For convenience during the experimentations, 450 images of 90 subjects with 5 images per subject selected randomly from each database except CASIA-IrisV2.0-device1. From CASIA-IrisV2.0-device1 database, 300 images of 60 classes are used for the experiments. MMU1 images provide less iris information than the selected iris databases.

A. D. Rahulkar and R. S. Holambe, *Iris Image Recognition*, 83
SpringerBriefs in Signal Processing, DOI: 10.1007/978-3-319-06767-4,
© The Author(s) 2014

References

1. Proenca H, Alexandre L UBIRIS: A noisy iris image database. http://www.iris.di.ubi.pt
2. Multimedia university (2004) MMU iris image database. http://pesona.mmu.edu.my/ccteo
3. CASIA iris image database. http://www.sinobiometrics.com/casiairis.htm
4. IITD iris image database. http://web.iitd.ac.in